# AVIATION

## *A WORLD OF GROWTH*

---

PROCEEDINGS OF THE 29TH INTERNATIONAL
AIR TRANSPORT CONFERENCE

---

August 19–22, 2007
Irving, Texas

SPONSORED BY
The Transportation & Development Institute (T&DI)
of the American Society of Civil Engineers

EDITED BY
Amiy Varma, Ph.D., P.E.

T&DI
Transportation and
Development Institute

Published by the American Society of Civil Engineers

Library of Congress Cataloging-in-Publication Data

International Air Transport Conference (29th : 2007 : Irving, Tex.)
Aviation, a world of growth : proceedings of the 29th International Air Transport Conference, August 19-22, 2007, Irving, Texas / sponsored by the Transportation & Development Institute (T&DI) of the American Society of Civil Engineers ; edited by Amiy Varma.
     p.cm.
Includes bibliographical references and index.
ISBN-13: 978-0-7844-0938-1
ISBN-10 0-7844-0938-2
  1. Aeronautics--Technological innovations--Congresses. 2. Airports--Congresses. 3. Aeronautics, Commercial--Congresses. I. Varma, Amiy. II. Title.

    TL505.I47   2007
    387.7'36--dc22                                  2007026129

American Society of Civil Engineers
1801 Alexander Bell Drive
Reston, Virginia, 20191-4400

www.pubs.asce.org

# Preface

Airlines and airports have managed to grow since 2001 despite low-fare competition, new security requirements, and increasing fuel prices. The growth in air transportation is no longer confined to the U.S. and Europe; it is a worldwide phenomenon now. There are many large-scale airport developments planned in Asia and Latin America. In the U.S. airports are preparing for new aircraft --A380, regional jets (RJs), and very light jets (VLJs)—which has brought new sets of challenges for airports. Managing growth and developing and financing the infrastructure to accommodate it are the challenges faced by federal, state, and local agencies, airport administrators, engineers and planners, program management teams, university researchers and consultants alike. The Transportation and Development Institute (T&DI) of ASCE has provided technical direction and support to the air transportation industry's growth and accomplishments and has played an important role in organizing numerous international air transportation conferences. This year's conference is the 29th International Air Transportation Conference (IATC).

The 29th International Air Transportation Conference provided a forum to address the planning and engineering of airports to meet the challenges of increased demand, newer aircraft and new security requirements. The Conference was designed to keep US and international civilian and military airport managers, designers and industry personnel abreast of the latest technology and innovative practices in the field of airport and airspace development. It provided opportunities to learn from leading experts worldwide as well as exchange ideas with peers. The technical program included technical tours, exhibits and general and concurrent sessions on various aspects of aviation development and technology.

The proceedings of the 29th International Air Transportation Conference (IATC) includes some of the papers that were presented at the conference. The conference theme was "Aviation – A World of Growth" and was held in Dallas-Fort Worth Metroplex, August 19-21, 2007. The conference brought together airport planners, designers, managers, researchers, industry professionals and development professionals from all over the world. The three day conference consisted of numerous conference sessions. Two of the plenary sessions were on lessons learned from airport development projects and airport sustainability. Two other plenary sessions focused on developments at DFW and Houston Airport Systems. The technical sessions were on emerging issues related to airfield design and construction, airport people movers and access, airport planning and safety, and airfield development case studies. There were three very informative and timely sessions dedicated to airport modeling, simulation and security related issues and developments. In addition, the attendees got an opportunity to have a technical tour of

DFW airport, where they got a first hand experience of and in-depth briefings on various development projects.

The Proceedings are not organized by session, but rather by groups which were related to particular theme. The papers are organized in five different categories—airport planning, safety, and management; airport modeling, analysis, and simulation; airspace and capacity innovations, airfield development case studies; and airfield design and construction.

The papers included in these proceedings went through a two step peer review process. The technical program subcommittee coordinated the peer review process. Larry Bauman and Geoff Baskir worked tirelessly to get commitments from authors and presenters. Jon Esslinger provided strategic input whenever there were questions. The Conference Steering Committee gratefully acknowledges the efforts of T&DI and ASCE staff, technical program subcommittee members, session organizers, reviewers, sponsors and to the many individuals and organizations who contributed to the success of the conference. The Conference Steering Committee members also thank the authors for their hard work in preparation of manuscripts and for cooperation and patience during the peer review process and the publication of proceedings. The conference benefited considerably from numerous presenters who shared their ideas, experiences and insights. All the presenters are thanked for their contribution. Finally, the Conference Steering Committee thanks the conference attendees who enhanced experience at the conference by quality exchange of knowledge and ideas.

**Amiy Varma, Ph.D., P.E., F.ASCE**
**Editor**

# Acknowledgments

## Conference Steering Committee

**Geoff Baskir**, C.M., AICP, M.ASCE, Conference Chair
**Larry Bauman**, C.M., P.E., Co-Chair, Technical Program Subcommittee
**William Fife**, P.E., M.ASCE, Co-Chair, Technical Program Subcommittee
**Amiy Varma**, Ph.D., P.E., AICP, M.ASCE, Conference Proceedings Editor
**Brian McKeehan**, P.E., A.M.ASCE, Chair, Sponsorships Subcommittee
**Perfecto Miguel Solis**, P.E., Chair, Local Coordination Subcommittee

## Technical Program Subcommittee

**Larry Bauman**, C.M., P.E., Co-Chair
**Gloria Bender**
**William Dunlay**, Ph.D., M.ASCE
**William Fife**, P.E., M.ASCE, Co-Chair
**Geoffrey Gosling**, Ph.D., M.ASCE
**Saleh Mumayiz**
**Shashi Nambisan**, Ph.D., P.E., M.ASCE
**Christopher Oswald**, A.M.ASCE
**Paul Schonfeld**, Ph.D., P.E., F.ASCE
**Richard Thuma**, P.E., M.ASCE
**Amiy Varma**, Ph.D., P.E., AICP, F.ASCE

## Reviewers

Several individuals who provided critical service and assisted technical program subcommittee with peer reviews of submitted manuscripts are gratefully acknowledged.

## T&DI/ASCE Staff

Andrea Baker
Donna Dickert
Jonathan Esslinger, P.E., F.ASCE
Varada Krishnaswamy

v.

# Contents

# China's Air Cargo Demand: Future Market Developments and Implications

## Ming-Cheng Wu[1], and Peter Morrell[2]

1. Research Student, Dept. of Air Transport, Cranfield University, Bedfordshire MK43 OAL, UK, PH: +44(0)1234 750111 ext 2242, FAX: +44(0)1234 752207; email: m.wu.2004@cranfield.ac.uk
2. Reader, Dept. of Air Transport, Cranfield University, Bedfordshire MK43 OAL, UK, PH: +44(0)1234 750111 ext 2242, FAX: +44(0)1234 752207; email: p.s.morrell@cranfield.ac.uk

## Abstract

China's economy has been rapidly growing with an average growth rate of 10% since 1990s. China's transformation, undoubtedly in response to its robust economy growth, has certainly become both so-called "the world's biggest market" and "the world's factory" in relation to its economy-related changes over the past decades. As a result, the perspective of China's air transport, in general, has been regarded as a fertile and competitive market in the current stage and even future years. The segment of air cargo services, in particular, is quite attractive and emerging industry in China's market recently, so that the competitors from all over the world are eager to scramble for part of China's air cargo markets. The purposes of the paper are to: (a) provide an analysis and the relationship between economic trends and air cargo traffic in China; and (b) forecast the aggregate demand of China's air cargo by using econometric method and extrapolate specific projections at China's major airports. In order to further realize the potential demand and competition of air cargo market between China, Hong Kong and Taiwan, we also provide an overall discussion on the interaction for the forecasting period.

**Review of Economic Development in China**

*Major Economic Stimulants*

Over the past 15 years (1991-2006), China's economic growth, unquestionably, had a brilliant development and will be resulting into transforming its economy into the world's strongest economic entity. China's GDP, in current prices, rose robustly from $371 billion in 1991 to $1934 billion in 2006 with a five-fold growth during the same period. Despite the impacts of the Asian currency crisis and the worldwide economic recession in 1997-1998 and 2001, respectively, China's economy still managed to sustain an average growth rate of 9.5%, as indicated by GDP growth index in Figure 1.

Another major stimulus to China's economic growth, the value of China's international trade, as shown in Figure 1, has risen approximately twelve-fold since 1990 with an average growth rate of 17% over the past 15 years. As of 2006, the estimated value of exports and imports reached 1594 billion USD, of which, exports was 875 billion USD and imports 719 billion USD.

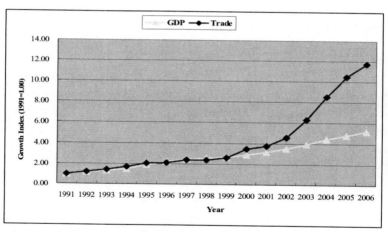

**Figure 1.China's GDP in Time Series Data (1991-2006) at Current Prices**
Source: National Bureau of Statistics (NBS) of China

*Economic Outlook*

China's economy-related development has been increasing since the 1990s, and many projections from recognized organizations and economic institutes indicate a very strong economic outlook for China through 2020. In order understand the implications of economic development, this paper analyzes for three scenarios-- low, baseline and high levels—based on assumptions regarding economic stimulants mentioned above.

With regard to China's GDP trend for the projection period, China's government expects to maintain a stable and reachable growth through 2020, projecting an average annual growth rate of 7.5%, and this growth rate is regarded as the baseline level. For high level it is assumed that China's average growth rate will be 9.5% through 2020. An average growth rate of 5.5 % is assumed for low level. Figure 2 shows the projects for three different scenarios.

With regard to China's trade growth through 2020, the three projections --low, baseline and high—are assumed to be an average growth rate of 7%, 10% and 13%, respectively. China's projection of trade growth is shown in Figure 3. Based on the time series data and projections for both GDP and trade, the relationship between China's air cargo traffic and economic activities will be explained in subsequent sections. In addition, forecast of China's air cargo demand in long-term takes into account the uncertainty of the economic factors.

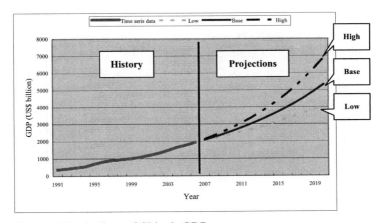

**Figure 2. Projections of China's GDP**
Source: National Bureau of Statistics (NBS) of China

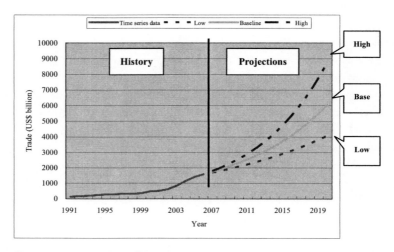

**Figure 3. Projections of China's Trade**

Source: National Bureau of Statistics (NBS) of China

## China's Air Cargo Throughput and Major Airports

### Emerging and Competitive Market

Air cargo throughput at China's airports, in terms of tonnage, rapidly increased from 787,000 in 1991 to 6,865,000 in 2006 with a nine-fold growth over the past 15 years. Air cargo traffic, in terms of tonnage, carried by Chinese airlines strongly increased from 455,000 in 1991 to 3,450,000 in 2006 with a over seven-fold growth. The total throughput at China's airports is closely related to the air cargo carried over the same period. Figure 4 compares the normalized throughput at China's airports with the normalized air cargo traffic carried by China's airlines (with 1991=1.00). Obviously the growth trend in both air cargo throughput and traffic carried for the past years paralleled with the similar increasing trend.

In 2006 approximately 7 million tons of air cargo was moved in China's airfreight routes, with 63% in domestic routes, 25% in international routes and 12% in regional routes. Due to lack of airfreight space and capacity of China's airlines, a high proportion of international air cargo was transported from/to China by foreign airlines in the past decade. Growth of air cargo demand and lack of China's airfreight capacity will continue in coming decade. An analysis of China's international air cargo flows for 2006 showed that 57 % of China's air cargo was moved by foreign airlines and only 43% was carried by Chinese airlines.

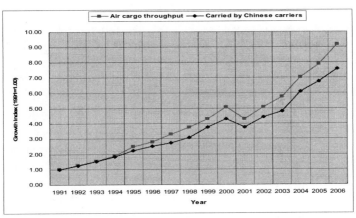

**Figure 4.Growth Indices for Air Cargo Throughput and Traffic (1991-2006)**
Source: Statistical Data on Civil Aviation, CAAC, China

Figure 5 indicates the growth indices related to China's air cargo, passenger, GDP and trade during 1994 to 2006 (1991=1.00). Growth index of China's air cargo was obviously higher than those for China's GDP and passenger over the past 10 years.

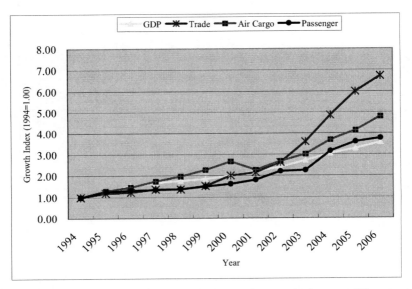

**Figure 5.Comparisons of Growth Indices for China's Air Cargo and Economy**
Source: 1. Statistical Data of Civil Aviation by CAAC, China.
2. National Bureau of Statistics (NBS) of China.

### China's Major Airport

According to the statistics from CAAC, the civil airports in China have increased from 100 in 1992 to 142 in 2006. China is positively planning to expand and enhance the facility and capacity at China's major airports in next five years. In general, all of the provincial capital cities have a major airport for the use by regional or national air transport. However, there are only a dozen of major airports along China's coastal regions (see Figure 6) that possess standard system facilities and enough transport capacity for national and international operations. In order to balance and accelerate the all-round development, the reform of China's airports has made great progress in recent years. Many positive reform measures, such as encouraging foreign capitals to construct airports and promote systems, localized management on airports, and opening up of air freedom rights, have brought great momentum for development of airfreight-related industries at China's airports.

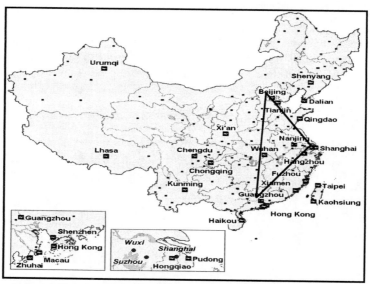

**Figure 6. China's Major and Secondary Airports**
Note: Only China's major airports are labeled and small blocks for secondary airports.

Air cargo throughput and average market share (in last decade) at China's major airports are shown in Table 1. The so-called "Big Triangle" of airports in China for air cargo is Shanghai, Beijing and Guangzhou whose air cargo throughput (both domestic and international) were 30%, 13% and 10% respectively in 2005. Thus, the three airports accounted for 53% of the total air cargo throughput. The remaining airports in China accounted for 47% of total air cargo throughput.

**Table 1. Air Cargo Throughput and Average Market Share**

| Major Airports | Cargo Throughput in 2006 (1000s of tons) | Average Market Share (1996-2006) |
|---|---|---|
| Shanghai (PVG/SHA) | 2159 | 29.0% |
| Beijing (PEK) | 1029 | 15.0% |
| Guangzhou (CAN) | 825 | 12.0% |
| Shenzhen (SZX) | 478 | 6.8% |
| Chengdu (CTU) | 268 | 4.5% |
| Kunming (KMG) | 208 | 3.4% |
| Hangzhou (HGH) | 177 | 2.5% |
| Xiamen (XMN) | 176 | 3.2% |
| Nanjing (NKG) | 151 | 3.8% |
| Wuhan (WUH) | 84 | 1.8% |
| Others | 1310 | 18.0% |
| All China | 6865 | 13.5% |
| Hong Kong (HKG) | 3609 | 9.2% |
| Taipei (TPE) | 1700 | 5.5% |

Source: 1. Statistical Data on Civil Aviation, CAAC, China.

2. Airports Council International.

## Air Cargo Demand and Specific Forecasts

### *Forecast Methodology*

In order to predict China's air cargo demand for the forecast period, the econometric approach with a log-log form was mainly applied to express the quantity of air cargo demand in this study. By using the analysis of the regression in the econometrics model, the forecast model of China's aggregate air cargo demand was established. The model made projections on the basis of the causal relationship between China's air cargo traffic and economic activities, which was obtained from time series data and economic projections for three different scenarios as discussed in aforementioned sections. Furthermore, the specific forecasts of air cargo throughput at China's major airports can be extrapolated and disaggregated by the average market share shown in Table 1.

Based on the methodology above, the regression model of China's aggregate demand is formulated, and a log-linear demand model is specified as follows:

$$Log\ (ACT) = a + b_1 * Log\ (GDP) + b_2 * Log\ (Trade) + b_3 * Log\ (FDI) \qquad (1)$$

Where, ACT = Air cargo traffic for China's aggregate Demand.

GDP = China's gross domestic product at current prices.

Trade = China's international trade at current prices.

FDI = China's foreign direct investment inflows.

Regression Model Summary:

| Variables | Constant | GDP | Trade | FDI | R | R Square | F Statistics |
|---|---|---|---|---|---|---|---|
| Coefficients | 2.413 | 1.035 | 0.422 | 0.363 | | | |
| t-statistics | 4.32 | 5.86 | 4.15 | 3.80 | 0925 | 0.856 | 376 |

$$Log\ (ACT) = 2.413 + 1.035 * Log\ (GDP) + 0.422 * Log\ (Trade) + 0.363 * Log\ (FDI)$$

### Forecast for China's Aggregate Demand

Applying the equation 1 and taking economic projections into account, there are three scenarios for China's aggregate demand, which result in alternative predictions of air cargo traffic (under "low", "most likely" and "high" scenarios) for the forecast horizon. China's aggregate air cargo demand and trends for three alternative scenarios are shown in Table 2 and illustrated graphically in Figure 7.

China's aggregate demand is expected to increase for "most likely" scenario from 7,948 thousand tons in 2007 to 26,253 thousand tons in 2020 with an average annual growth rate of 9.6%; an increase by 3.3 times over the forecast period. Compared to the average growth rate shown by Boeing's and Airbus's predictions, this figure at the "most likely" scenario is conservatively lower than both Boeing's (Boeing, 2007) and Airbus's (Airbus, 2006) figures of 10.8% and 10.2%, respectively.

**Table 2. China's Aggregate Demand for Air Cargo Throughput**

| Historical (1000s of tons) | | Forecast Demand (1000 of tons) | | | |
|---|---|---|---|---|---|
| Year | Traffic | Year | Low | Most Likely | High |
| 1993 | 1230 | 2007 | 7398 | 7615 | 7835 |
| 1994 | 1500 | 2008 | 7924 | 8349 | 8791 |
| 1995 | 1950 | 2009 | 8480 | 9154 | 9864 |
| 1996 | 2200 | 2010 | 9079 | 10034 | 11067 |
| 1997 | 2600 | 2011 | 9720 | 11002 | 12424 |
| 1998 | 2950 | 2012 | 10405 | 12059 | 13940 |
| 1999 | 3405 | 2013 | 11135 | 13221 | 15644 |
| 2000 | 4002 | 2014 | 11922 | 14490 | 17551 |
| 2001 | 3400 | 2015 | 12762 | 15887 | 19696 |
| 2002 | 4020 | 2016 | 13666 | 17415 | 22103 |
| 2003 | 4520 | 2017 | 14626 | 19092 | 24801 |
| 2004 | 5526 | 2018 | 15659 | 20928 | 27833 |
| 2005 | 6006 | 2019 | 16762 | 22946 | 31233 |
| 2006 | 6865 | 2020 | 17948 | 25153 | 35049 |

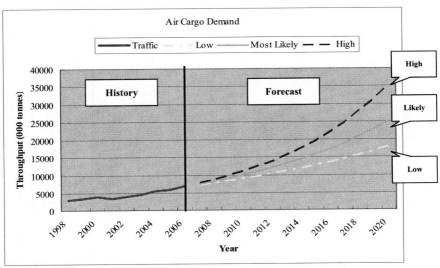

**Figure 7. China's Aggregate Air Cargo Demand for Alternative Scenarios**

### Projections for China's Specific Airports

Based on the historical traffic in Table 1, air cargo throughput at China's major airports was compared with those at Hong Kong and Taipei airport. For "most likely" scenario air cargo projections for airports in China are provided in Table 3 and also shown in Figure 8.

As shown in Table 3, air cargo (for both domestic and international routes) at Shanghai airport is projected to increase by 5035 thousand tons; from 2159 thousand tons in 2005 to 7294 thousand tones in 2020 representing a three and half-fold growth. In addition, air cargo demand at Shanghai airport will be predicted to outpace the throughput at Hong Kong by 2017 (see Figure 8). These predictions take into account many constraints limiting the throughput in Shanghai, such as airport capacity and facilities, cargo demand and services and other ground transportation facility and services.

**Table 3. Air Cargo Projections at China Airports for the "Most Likely" Scenario**

| Major Airports | Historical Traffic (000 tonnes) | | | Forecast Demand (000 tonnes) | | | |
|---|---|---|---|---|---|---|---|
| | 2004 | 2005 | 2006 | 2007 | 2010 | 2015 | 2020 |
| Shanghai (PVG/SHA) | 1636 | 1917 | 2159 | 2208 | 2910 | 4607 | 7294 |
| Beijing (PEK) | 669 | 782 | 1029 | 1142 | 1505 | 2383 | 3773 |
| Guangzhou (CAN) | 507 | 595 | 825 | 914 | 1204 | 1906 | 3018 |
| Shenzhen (SZX) | 423 | 464 | 478 | 518 | 682 | 1080 | 1710 |
| Chengdu (CTU) | 213 | 251 | 268 | 343 | 452 | 715 | 1132 |
| Kunming (KMG) | 171 | 197 | 208 | 259 | 341 | 540 | 855 |
| Hangzhou (HGH) | 128 | 166 | 177 | 190 | 251 | 397 | 629 |
| Xiamen (XMN) | 142 | 159 | 176 | 244 | 321 | 508 | 805 |
| Nanjing (NKG) | 118 | 139 | 151 | 289 | 381 | 604 | 956 |
| Wuhan (WUH) | 62 | 70 | 84 | 137 | 181 | 286 | 453 |
| Others | 1157 | 1170 | 1310 | 1371 | 1806 | 2806 | 4528 |
| All China | 5526 | 6210 | 6865 | 7615 | 10034 | 15887 | 25153 |
| Hong Kong (HKG) | 3119 | 3433 | 3609 | 3771 | 4279 | 5176 | 6100 |
| Taipei (TPE) | 1701 | 1745 | 1700 | 1790 | 2150 | 2918 | 3916 |

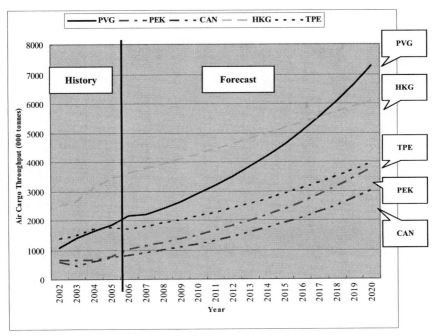

**Figure 8. Air Cargo Projections for China, Hong Kong and Taipei**
Source: Historical numbers from
        1. Civil Aeronautics Administration (CAA), Taiwan
        2. http://www.cksairport.gov.tw
        3. http://www.hongkongairport.com

From the economic stimulus, a major driver for Shanghai air cargo growth is the strength of the economy in Yangtze River Delta. Success in binding the Delta-based companies and industries to the Shanghai hub, by means of high-speed transport and marine/river links, will be a key success factor for air cargo hub in the long term. Other factors influencing Shanghai's future growth as an air cargo hub will be: (1) successful development of domestic and international air services links via Shanghai; (2) attracting investment and businesses around Shanghai and the Yangtze region; (3) streamlining customs and import-export procedure with an information-technology infrastructure.

## Conclusions

China's aggregate demand for air cargo throughput is predicted to have an increase of 18000 thousand tons from 6865 thousand tons in 2006 to 25153 thousand tones in 2020 under "most likely" scenario; accounting for an average growth rate of 9.6%

annually. For the specific projections for China's major airports, air cargo demand of the "Big Triangle" in Shanghai, Beijing and Guangzhou will still be the three largest airports by 2020, and the total of their air cargo throughput will continue to have a market share of approximately 50% for both China's domestic and international air cargo market during the forecasting period. Shanghai, in addition, will even outpace Hong Kong's throughput by 2017.

Shanghai's growth of air cargo traffic has been stronger than any other airport in China over the past decade, and as such Shanghai has been regarded as the most important international air cargo hub by China for the coming decade. As to the air cargo projection in Taipei, it will benefit Taiwan's air cargo market and attract some part of traffic from Hong Kong's air cargo transshipment if there is opening of the direct air links between Taiwan and China. For the traffic analysis, a large proportion of Shanghai's air cargo activity has been "origin-destination" flow, with cargo either originating from or destined to Shanghai and its adjacent provinces.

In addition, the implications for system facilities and airport capacity at China's airports are very significant to air cargo development, and some outdated systems at major airports should be enhanced and replaced in order to reach the standard level for the rapidly increasing market in China. CAAC and local governments will continue expanding and constructing several airports in China's metropolitan areas for the balanced air transport development according to China's 11th five-year plan.

## References

Airbus Industrial Group, Global Market Forecast 2006-2025, October 2006. http://www.airbus.com

Airports Council International, Global Traffic Forecast 2006-2025 and Data Center. http://www.airports.org

Boeing Commercial Airplanes Group, World Air Cargo Forecast 2006-2007, http://www.boeing.com

Civil Aeronautics Administration (CAA), Taiwan. http://www.caa.gov.tw

CAAC, Statistical Data on Civil Aviation of China, 1996~2004.

General Administration of Civil Aviation of China (CAAC), Statistics Center, http://www.caac.gov.cn

Hong Kong International Airport, Annual Air Cargo Statistics 2005, http://www.hongkongairport.com

National Bureau of Statistics (NBS), China. http://www.stats.gov.cn

Taiwan Taoyuan International Airport, Air Cargo Statistics and Data Center, http://www.cksairport.gov.tw

# A Personal Rapid Transit/Airport Automated People Mover Comparison

Peter J. Muller, P.E.[1]

[1]PRT Consulting, Inc., 1340 Deerpath Trail, Suite 200, Franktown, CO 80116
PH (303) 532-1855; FAX (303) 309-1913; Email: pmuller@prtconsulting.com

## Abstract

Airport automated people movers (AAPM) typically consist of driverless trains with up to about four cars each capable of carrying 20 to 100 passengers who are mostly standing. They have been successfully used for surface transportation in airports for over thirty years. A new category of automated people mover called personal rapid transit (PRT) is being implemented at London's Heathrow International Airport. Although the Heathrow system will replace shuttle buses, it may be more pertinent to examine the differences between PRT and traditional AAPM. PRT uses small (3 to 4 passenger) vehicles (transportation pods or T-Pods) to automatically transport passengers and their luggage non-stop to their destinations along designated guideways. Trips are typically on-demand and T-Pods are often waiting at stations prior to the arrival of passengers. The resulting short wait and trip times combine with seated travel to provide an exceptionally high level of service. This paper compares AAPM systems to PRT systems similar to the type being installed at Heathrow Airport. Items compared include infrastructure items such as stations, guideways and tunnels; level of service items such as waiting, standing and trip times; cost items such as capital and operating costs; as well as safety and security issues. The paper discusses PRT viability and concludes with a brief discussion of the ability of PRT to facilitate solutions to common airport issues such as in-concourse transportation and curbside congestion. PRT is found to have many advantages over AAPM for transporting passengers and their luggage on airports. It is suggested that PRT alternatives should be included in airport planning projects.

## Introduction

The first personal rapid transit (PRT) system came into service at West Virginia University in Morgantown over thirty years ago, a few years after the first airport automated people mover (AAPM) began operation at the Tampa International Airport. The key difference between these two systems was that the Morgantown system could operate only when needed (on-demand) and bypass vehicles stopped in stations thereby taking its passengers directly to their destinations non-stop. Another difference was that the Morgantown system suffered considerable teething problems resulting in other proposed PRT systems being cancelled and further PRT

14

development languishing for about three decades. Modern PRT systems that almost all use much smaller vehicles than either the Morgantown PRT system or AAPMs are now rapidly emerging. The operating characteristics of these very small systems are quite different than those of conventional AAPMs and this paper is intended to provide an overview of how modern PRT systems compare to conventional AAPMs.

## PRT Characteristics

While the Morgantown system is called a PRT system, it does not meet the common definition of PRT and is more correctly classified as group rapid transit (GRT). This paper is focused on a definition of PRT that, in the author's opinion, is best suited for airport applications. This definition is outlined below and is similar to that provided by the Advanced Transit Association (2003):

- Small T-Pods (4 passengers plus their luggage)
  - o Passengers all traveling together to same ultimate destination (little or no shared rides)
- On-demand, non-stop service
  - o Little or no waiting
- Operates inside buildings
- T-Pods are typically constrained to guideways
- Guideways are usually separated from other traffic
- Guideways can be at-grade, elevated or below grade
- Small turning radius (<20')

PRT vendors are presently providing (or planning to provide) three different types of PRT systems – open guideway; captive bogey, and; suspended, as illustrated in Figures 1, 2 and 3. It is likely that the open guideway systems will prove to be more common in airport applications primarily because of their small turning radius capability. This paper compares open-guideway PRT systems with conventional AAPMs.

## Infrastructure

This section compares PRT and AAPM infrastructure in terms of the requirements for elevated and at-grade structures, tunnels and stations.

### Guideways

The small size of PRT vehicles (see Figure 4) results in small-scale infrastructure being required. However, it can also result in limited capacity. Low-speed (less than 25 m.p.h. (40 km/h)) PRT systems can safely operate at headways as low as 2 seconds (PRT Consulting, Inc). While it is possible that lower headways will prove safe for PRT systems, this figure is used in this paper. Four-seat T-Pods at 2 second headways offer a maximum theoretical capacity of 7,200 passengers per hour per direction (pphpd). This compares to the maximum theoretical capacity of a typical AAPM with trains consisting of four 100-passenger cars and operating at 90 second headways of 16,000 pphpd. Thus one AAPM guideway could have more than twice the capacity of one PRT guideway. This suggests that, for guideway costs to be comparable, PRT guideways should cost about one half of AAPM guideways.

Figure 1. ULTra's At-grade Open Guideway.

Figure 2. Postech's Captive-bogey Guideway.

Figure 3. JPod's suspended system

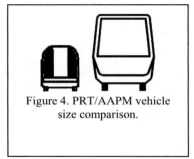

Figure 4. PRT/AAPM vehicle size comparison.

However, this is not always the case as can be demonstrated by considering systems where the desired theoretical capacity is 7,000 pphpd or 20,000 pphpd. In the former case the PRT guideway could have the same costs as the AAPM guideway; in the latter case the PRT guideway should cost two thirds the cost of the AAPM guideway to be comparable.

Another complicating issue when comparing guideway costs is that AAPM systems are usually laid out as two-directional guideways serving a corridor. PRT systems can be laid out in this manner too but can sometimes be more beneficial as one-way guideways which can serve a wider area but then may require additional inter-connecting loops resulting in more total guideway length. To overcome this type of difficulty in comparing dissimilar systems, it is sometimes desirable to compare total system costs on a per-station basis.

**Elevated**

PRT elevated guideways need to carry a live load of less than 10 tons per span while AAPM elevated guideway spans need to support about four times this weight. PRT column loads are approximately 10 to 12% of typical AAPM column loads (Kerr, 2005). Kerr states that an elevated PRT structure has a significantly lower loading than a footbridge which must accept crowding loads.

**At Grade**

The at-grade requirements for an open-guideway PRT system are not much more than that for a pedestrian sidewalk. Typically the guideway can consist of a seven foot (2.1m) wide pavement with eighteen inch (450mm) high sidewalls, all consisting of six inch (150mm) thick concrete and typically placed on about six inches (150mm) of gravel base. Where AAPMs run at grade they are typically supported by two, two foot by two foot (600mm x 600mm) concrete tracks on a supporting foundation varying in dimension according to the support capabilities of the subgrade soils. Comparing just the amount of concrete described above, the PRT guideway requires about 62% of the AAPM guideway.

**Below Grade**

Two PRT guideways will fit into a tunnel of half the cross-sectional area required for one AAPM guideway (Muller, 2005). Comparing PRT capacity to road capacity, a 200 square foot ($19m^2$) PRT tunnel could exceed the passenger-carrying capacity of a 950 square foot ($90m^2$) vehicular tunnel (Lowson 2005).

***Stations***

AAPM station lengths are typically governed by the maximum train length while the width is designed to accommodate the AAPM tracks plus all of the people boarding and deboarding. This results in a fairly wide platform since there could theoretically be a trainload of people boarding and another deboarding at one time at one station. Typically the design peak number is somewhat less than this theoretical maximum number but nonetheless can be quite large. The AAPM station at Denver International Airport (DIA) Concourse B is 177 feet long by 32 feet wide (excluding the tracks) and has an area of 5,667 square feet ($527m^2$). It has an estimated capacity of about 4,000 passengers per hour.

Using a 30 second dwell time, one PRT station bay can serve about 120 T-Pods per hour. If the average T-Pod occupancy is 2.0 (some ride sharing is likely during peak periods in an airport), this means each station bay can accommodate 240 passengers per hour per direction for a total of 480 passengers. Thus 9 bays would be needed to match the AAPM number for DIA concourse B. A 9-bay PRT station has an area of about 4,340 square feet ($403m^2$) (excluding the tracks except at each vehicle bay) which is 76% of the area of the AAPM station.

In practice, it may make more sense for a PRT system to have more stations with fewer bays, thus reducing walking distances. Such an arrangement will usually increase the relative PRT station area.

One of the major differences between AAPM and PRT stations results from the flexibility of the smaller PRT systems. Their ability to accommodate tight radii and steep gradients makes it possible for PRT stations to be accommodated within concourses and potentially at or close to grade. This offers the potential of easier use resulting in higher patronage. It also could result in the elimination, or reduction in number, of escalators and elevators. Figure 5. depicts a station designed to fit within Concourse B at DIA. It has a footprint smaller than that of the existing moving sidewalks.

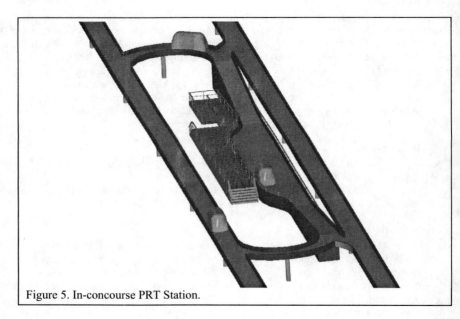

Figure 5. In-concourse PRT Station.

## Costs

Unfortunately there is not much good data available with which to compare either capital or operating costs of PRT with AAPMs. PRT is too new for much data to have accumulated and AAPM costs are often presented in a way that makes it difficult to determine just what elements (such as stations) are included in the costs.

### Capital Costs

The only known source of modern PRT capital costs based on a construction contract is the ULTra project at Heathrow International Airport. The capital cost including guideways, stations, vehicles and operating system but excluding column footings is reported to be less than US$10 million per one-way mile. This cost is for a system that is not expected to have a high demand and is thus probably on the low end. This type of open-guideway PRT system can be expected to cost between US$10 and US$15 million per one-way mile (Advanced Transport Systems Ltd.).

Kerr (2005) provides AAPM capital costs of US$24 to 75 million per one-way mile. This is approximately 2.4 to 5 times the PRT costs quoted above and indicates that PRT systems will usually cost less than AAPM systems for the same capacity. This seems to confirm the results reported by Muller (2005) in a study comparing actual AAPM capital costs at Denver International Airport with estimates provided by three different PRT vendors. This study found the PRT capital cost to be 35% of the AAPM capital costs.

*Operating Costs*

AAPMs are typically installed in high capacity situations and therefore have relatively low (less than $1.00) operating costs per passenger. Modern PRT systems are expected to have even lower operating costs but this has not yet been proven. The Morgantown PRT system has operating costs of about $1.50 per passenger.

**Level of Service**

PRT level of service is designed to better match that of an automobile in uncongested conditions than that of any conventional form of transit including AAPMs. Service is on-demand with little or no waiting. Passengers are taken directly to their destinations with no stopping and in seated comfort. Way finding is simplified because the only knowledge needed is the ultimate destination – the system will find the best route. Passengers are transported in privacy with little or no need to share rides with strangers. There is little or no need for the trip to be punctuated by a public address system.

While PRT travel speeds are likely to be relatively slow initially (25 m.p.h. or 40 k.p.h.), the total trip times are likely to be less than AAPM trip times because of the reduced waiting times and the elimination of intermediate stops. Muller (2005) found PRT trip times to be 45% of AAPM trip times. In addition, PRT systems are likely to have more stations resulting in reduced walking distances and times.

**Safety and Security**

PRT systems are expected to be significantly safer than conventional transit systems and to match the safety record of AAPMs. The Morgantown PRT system has completed over 110 million injury-free passenger miles (Muller, 2007) providing evidence of the safety of PRT operating concepts.

Small vehicles providing on-demand service at small stations result in a lack of crowding, which in turn means that PRT systems do not present likely terrorist targets. In addition, PRT systems deliver a steady stream of traffic which could facilitate security screening. Future PRT systems could be equipped with on-board check-in kiosks which could facilitate airline transactions as well as the gathering of security pre-screening data. Ultimately it may be possible to prescreen passengers and their bags for undesirable substances while they are traveling in the PRT vehicles.

**PRT Viability**

The Morgantown PRT system has proven the viability of the PRT concept. The system can and does operate in an on-demand, PRT mode (as well as in other modes). Since the manufacturer (Boeing) no longer provides PRT systems, there are currently no PRT suppliers with proven viability. Until recently, the only PRT suppliers (except for 2getthere, a Dutch company with associations with larger companies) were small, relatively under-capitalized companies. This is rapidly changing.

When BAA selected the ULTra system for Heathrow Airport, they liked it so much they decided to purchase stock in Advanced Transport Systems, Ltd., the supplier. The Korean steel company, Posco (one of the world's largest) has formed a subsidiary called Vectus that has constructed a significant test track in Sweden and is aggressively developing a PRT system. Thus PRT suppliers now include a number that are backed by billion dollar companies.

Nonetheless, there are presently no modern PRT systems in public operation. 2getthere has a GRT system operating in Holland and has had experience with other PRT-like systems operating in the public domain. The first modern PRT system is under construction at Heathrow Airport and is scheduled to come into public service in 2008. Other PRT deployments are anticipated to follow rapidly.

**Potential Airport Design and Operational Impacts**

This paper indicates that PRT could provide better service than AAPM at lower costs in many airport applications. However, PRT has the potential to be more than just an improved, lower cost version of AAPM. Its flexibility, low cost and high level of service combine to potentially allow it to change the way airports are designed and operated.

A PRT system replacing the airport shuttle bus system serving parking lots and rental car facilities could reduce curbside congestion and the need for consolidated rental car facilities. Such a system could completely avoid the curbside by bringing passengers right into a mezzanine level of the terminal as depicted in Figure 6.

AAPMs allowed airports to have remote concourses connected to the main terminal by the underground people mover. However, these concourses have become very long (up to one mile (1.6km)) and can involve considerable walking distances. PRT systems have the flexibility to rise up into the concourses and deliver passengers to within a short distance of their gate rather than to a central underground station. This could allow seated travel from the terminal to the gate. It could also allow long concourses to be divided into multiple shorter concourses. These shorter concourses could be quite a significant distance from each other and/or from the main terminal potentially allowing additional flexibility in airport layout planning.

In the future, on-board airline functions and security screening (facilitated by each T-Pod only carrying small groups all traveling to the same destination) may change the required functionality of the terminal building and allow just-in-time passenger delivery to the gate. This concept has the potential to consolidate waiting in a consolidated concession area thus helping to improve the airport's bottom line.

Figure 6. Rendering of a PRT system inside the DIA main terminal. While the system looks large in the foreground, observing the return guideway in the background provides an appreciation of its small scale.

## Conclusions and Recommendations

Table 1 summarizes the findings of this paper. For each category PRT and AAPMs are rated on a scale of 1 to 5 with 1 being poor and 5 being excellent. Note that the ratings are intended only to highlight the differences between these two systems and thus a system rating poorly relative to the other in any one item may still be quite acceptable in that item.

**Table 1. Summary of Results. 1 = Poor, 5 = Excellent.**

| Item | AAPM | PRT |
|---|---|---|
| Elevated guideways | 2 | 4 |
| At grade guideways | 2 | 3 |
| Below grade guideways | 2 | 4 |
| Stations | 2 | 3 |
| Capital costs | 1 | 3 |
| Operating costs | 3 | 4 |
| Level of Service | 2 | 5 |
| Safety | 4 | 4 |
| Security | 2 | 4 |
| Viability | 4 | 3 |
| Flexibility | 2 | 4 |
| Potential side benefits | 2 | 5 |

There is little doubt that modern PRT systems are coming and will soon be operational at airports. The opportunities they bring for improving airport functionality and revenue as well as passenger level of service are such that airport planners should be taking PRT into account now as they develop their plans for improving airports around the world.

## References

Advanced Transit Association, *Personal Automated Transportation*, January, 2003, http://www.advancedtransit.org/doc.aspx?id=1015

Advanced Transport Systems Ltd., http://www.atsltd.co.uk/prt/faq/

Kerr, A.D. et al, *Infrastructure Cost Comparisons for PRT and APM*, ASCE APM05 Special Sessions on PRT, 2005.

Lowson M.V., *PRT for Airport Applications*, Transportation Research Record, Journal of the Transportation Research Board No 1930 2005 pp 99-106

Muller, P. J., Allee, W., *Personal Rapid Transit, an Airport Panacea?* Transportation Research Board Paper No. 05-0599, January, 2005

Muller, P. J., et al, *Personal Rapid Transit Safety and Security on a University Campus,* Transportation Research Board Paper No. 07-0907, January, 2007

PRT Consulting, Inc. *Analysis of Capital Costs,* http://www.prtcons.com/Informative.htm

# Implementing a Perimeter Taxiway at Dallas/Fort Worth International Airport: Evaluation of Operating Policy Impacts

S. D. Satyamurti, Ph.D., P.E.[1] and Stephen P. Mattingly Ph.D. [1]

[1]Department of Civil and Environmental Engineering, University of Texas at Arlington, Box 19308, Arlington, TX 76019-0308; PH (817) 272-2859; FAX (817) 272-2630; email: satyamurti@yahoo.com, mattingly@uta.edu.

**Abstract**

A perimeter taxiway (PT) or end-around taxiway (EAT) operation is a new concept being developed at several airports with parallel runways around the United States to reduce the number of active departure runway crossings during peak periods. PTs will enhance capacity by permitting uninterrupted, safe, continuous takeoffs and landings and improve safety by reducing the likelihood of runway incursions. For this research, the Dallas/Fort Worth International Airport's (DFW) proposed PT operations are considered for analysis and evaluation. This concept is tested using Visual SIMMOD simulation modeling software where the input data is based on actual historical DFW flight data and preliminary design documents. The PT operational analysis is based on the simulation results using forecast air traffic data for 2010. The operations analysis, and the standard taxiway procedures and guidelines developed based on the simulation gives an indication of the types of operational issues that may develop at DFW and other airports after PT implementation.

## Introduction

### Problem definition

The air traffic at towered airports throughout the United Sates (US) is growing at a steady rate in line with the growth in economy and population (FAA 2004a). The global market demand for commodities and services has added a new dimension to the concept of travel. Far East and Asian countries have become leaders in manufacturing, which has resulted in the movement of people, raw materials, and finished goods to destinations around the world (Airbus 2005). The increase in traffic occurs simultaneously with the introduction of new long-range and short-range aircraft to carry passengers on international and domestic routes (Boeing 2005). There is no longer a peak traffic period at major airports like Dallas Fort Worth International Airport (DFW), O'Hare International Airport (ORD), Los Angeles International Airport (LAX), Atlanta Hartsfield International Airport (ATL) and San

23

Francisco International Airport (SFO) (FAA 2005). At these airports, airlines have rescheduled their flights over the entire day instead of clustering arrival and departure slots together in the morning or afternoon. This has helped to decrease severe delays and has greatly reduced the communication requirements and workload for air traffic controllers. The net effect is better use of gates and baggage handling facilities at these airports (FAA 2004b). The parallel runway operations at towered airports cause aircraft to wait before they cross the departure runway to reach the terminal gates. The waiting time is increasing and a solution to eliminate this wait time, which will simultaneously make operations safer, save fuel and improve overall gate usage and increase facility utilization while maintaining on time arrivals and departures (Davis 2002).

The planned addition of the PT at DFW will increase the safety and reduce operational constraints during peak periods (Davis 2002). While a PT provides operational benefits to an airport by reducing the waiting time for aircraft to cross the active departure runway after arrival, this paper considers new operational constraints that may result from the PT implementation. The objective is the smooth movement of aircraft from the arrival runway to the gate with minimum communication with Ground Controllers. With a PT, the aircraft on the departure runway need not wait for the arrival aircraft to cross, which allows continuous departures on the dedicated departure runways 17R and 18L. This paper investigates the operating policy impacts by simulating the proposed DFW PT using Visual SIMMOD (VS) software and identifying the PT operational benefits and constraints on all four quadrants.

### Current Status at DFW

The FAA approved the design and construction of the SE quadrant PT in September 2006. The PT layout used in this research study matched the final designs and construction drawings (AOSC 2005). The PT centerline was set at 2,650 feet from the end of the two north-south parallel runways, 17C and 17R. The contract was awarded on 10 October 2006 at a cost of $66.7 million (FAA funding 75%) with completion expected in the fall of 2008 (Associated Construction Publications 2006). Mr. Jim Crites, Executive VP of Operations at DFW had this to say, "This is a win-win-win situation. By installing a perimeter taxiway system, we will be providing a better and safer operating environment for both pilots and air traffic controllers who devote themselves to providing a safe and efficient operating environment. The system will also provide the traveling public with greater efficiency and a small amount of delay on the ground, getting them off the gate or to their gates faster than ever before" (Associated Construction Publications 2006). From a discussion with the FAA/ATC staff at DFW, it was mentioned that each flight will be monitored with regard to their assigned terminals and a decision would be made at that time to permit active runway crossing or to direct them to taxi on the PT to reach the terminals or vice versa to access the departure runways from the gates. No doubt this would increase the potential of runway incursions when the operating guidelines are modified from PT to non-PT during flight operations. Therefore, these operational changes would require due diligence and constant communication to avoid conflict and runway incursion at DFW.

### DFW Configuration

The current configuration at DFW shown in Figure 1, requires that aircraft arriving on the main arrival runways 13R, 18R/36L, 17C/35C, 17L/35R and 31R cross the main dedicated inboard departure runways 18L/36R and 17R/35L to get to the terminal area. Depending on the direction of air traffic flow and whether or not aircraft are arriving on the three outboard runways, many arriving aircraft have to cross two runways (both arrival and departure) to get to the terminal area. Similarly, the departing aircraft from the terminals, east and west cargo aprons have to cross a departure or arrival runway depending on the assigned departure runways 13L, 17R/35L, 18L/36R and 31L. Runways 18R/36L and 17C/35C are also used for departures depending on the flight destination or arrival frequency. It is estimated that on average DFW experiences over 1,700 runway crossings daily (FAA Runway Safety 2003).

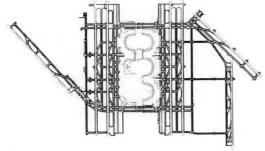

**Figure 1. DFW Layout in 2004**

At DFW, there are five terminal buildings, A, B, C, D and E, as well as, general aviation, east and west cargo facilities. Under existing operations, the Local Air Traffic Controller conducts all runway crossings before releasing the aircraft to the Ground Controller. This situation increases the Local Controller's workload and creates radio frequency congestion. During major arrival and/or departure periods, trade offs in airfield efficiency have to be made to safely balance all operations (Erway 2003).

Leigh Fisher Associates (LFA 1996) performed the first study on PT operations and considered several configurations and estimated the runway crossing delay. In 2002, Davis (Davis 2002) conducted a detailed study of implementing a PT system at DFW. Davis (2003) analyzed the obstruction free zone (OFZ) criteria and proposed that the PT should be centered about 2650 ft from end of the north-south parallel runways at DFW. In 2003, a demonstration was conducted in a flight simulator at the NASA's Ames Lab at the Moffet field in California (Buondonno 2003). From these studies, it was revealed that the PT allows the aircraft to go around the active departure runway without crossing the runway to reach the terminal buildings, thus increasing safety of operations and an increase in departures. There was reduction in communication between the cockpit and the tower during the PT operation. This allows the flow of arrival aircraft to reach the terminal without having to wait for clearance from the Local Controller or Ground Controller to cross the departure runway. This will greatly increase the efficiency of operations, reduce runway incursions and considerably decrease communications between the Ground Controllers and the cockpit (Buondonno 2003).

This balancing partially consists of controllers delaying departing aircraft so that arriving aircraft can cross the departure runways to get to the terminal area. Because arrivals stack up at the various runway-crossing points, the Local Controller must "gap" departures to allow these crossings to occur. These situations are most evident during the peak traffic times (Erway 2003). In an effort to improve safety and airfield efficiency by reducing the number of active runway crossings (with the added benefit of reducing runway incursion potential and reducing arrival and departure delays), a PT concept was proposed. The concept includes new PTs on the East and West sides of the airport.

The actual operations for 2004 are 718,270 and the forecast is 845,502. The PT system layout is shown in Figure 2, which will enable the arrival aircraft to taxi without waiting for clearance from Ground Controllers. The aircraft may have to taxi a longer distance to reach the gate, but results in a significant reduction in communications with controllers on the ground control tower and the elimination of runway crossing delay. The aircraft will be able to move in an orderly queue, thus permitting continuous takeoffs on the departure runway without the risk of runway incursions. The aircraft spacing on the takeoff runway is based on the allowable distance between aircraft as specified in FAA standards and guidelines.

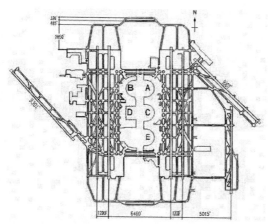

**Figure 2. Perimeter Taxiway Layouts**

Based on current operations, the departures from the primary departure runways 17R/35L and 18L/36R are expected to be steady based on enroute weather, traffic and conditions at the destination cities. It will help the Ground Controllers to schedule departures without concern for arriving flights during the PT operations. This departure procedure is replicated in the VS simulation for this research.

## Simulation Issues

### *Demand for Air Travel*

The Dallas Fort Worth Metroplex is experiencing rapid population growth with the arrival of new industries, support services and financial institutions {NCTCOG 2003]. Air traffic at DFW is expected to increase in the coming years, which will certainly increase the runway crossings, and the associated delay to both incoming aircraft waiting to cross and to departing aircraft [Buondonno 2003]. After completing a detailed review of FAA forecasts for DFW, this research estimated the traffic flow at DFW in year 2010.

There is a distinct possibility that the air cargo traffic may increase over the forecast years, which will depend on availability of additional cargo apron and facilities at the airport. At present, there are many gates not in full use at Terminals D and E. Therefore, the airport will be able to handle additional passenger flights without substantial investment on new terminal buildings in the near future. The analysis of forecast growth in passenger traffic from different sources yields an anticipated growth rate of 3.5% to 4% through year 2030. The terminal area forecast from FAA predicts an increase of 3.5% per year for air traffic operations at DFW [FAA 2005]. For this paper, the air traffic operations at DFW was forecast to increase at the rate of 3.5% per year from 2004 to 2010 in line with the overall growth projected by the FAA for the US airline industry. The 2004 actual flight schedule data was obtained from the DFW database for every day of the year for all scheduled flights. The aircraft type information for each airline and the runway used by each flight at DFW was obtained from the flight tracks/operations database of the DFW Environmental Affairs Department (EAD) and the Official Airline Guide (OAG).

A detailed analysis of the air traffic data at DFW, which is shown in Table 1, indicates that the highest daily operations (2,477) occurred on July 22, 2004. The detailed arrival and departure schedule obtained from the DFW scheduling department gave the flight number, arrival and departure time, origin, and destination cities, and the gate assignment for each flight. The flight schedule timetable received from the OAG for July 2004 gave information on flight schedule for all airlines serving DFW with flight number, scheduled arrival and departure time, origin and destination cities and the aircraft type used. The data files were reviewed, analyzed, and consolidated into one composite file for VS input.

### *DFW Operations*

The "Runway Use Plan" [DFW Airport 1996] document is the basic document for assigning arrivals and departures in VS. The direction of flow is indicated in the runway use diagram. The flight operations at DFW include scheduled flights by air carriers, air cargo, military and air taxi. The aircraft type in use on each flight is obtained from DFW EAD database, FAA/APO Aviation System Performance Metrics (ASPM) reports, and the Official Airline Guide (OAG)

## Table 1. 2004 DFW Actual Operations Data

| Month | Total | Maximum | Minimum | Mean | Range |
|-------|-------|---------|---------|------|-------|
| JAN | 68,425 | 2381 | 1950 | 2207 | 431 |
| FEB | 64,039 | 2358 | 1653 | 2208 | 705 |
| MAR | 69,317 | 2384 | 1647 | 2236 | 737 |
| APR | 67,961 | 2421 | 1981 | 2265 | 440 |
| MAY | 69,861 | 2405 | 1976 | 2254 | 429 |
| JUN | 68,511 | 2434 | 2038 | 2284 | 396 |
| JUL | 70,571 | 2477 | 1837 | 2276 | 640 |
| AUG | 70,650 | 2421 | 1931 | 2279 | 490 |
| SEP | 66,113 | 2408 | 1737 | 2203 | 671 |
| OCT | 67,714 | 2394 | 2147 | 2184 | 1201 |
| NOV | 64,930 | 2361 | 1665 | 2164 | 696 |
| DEC | 65,450 | 2275 | 1719 | 2111 | 556 |
| TOTAL | 813,542 | | | | |

Source: www.apo.data.faa.gov accessed on 8-23-06

### Validation of VS Performance

The FAA established efficiency criteria [Wine 2005] were used to compute the efficiency of operations at DFW and compare the observed DFW and simulated efficiencies. The results for the sixteen simulated scenarios are shown in Table 2. Three simulated dates, 7-22-04 (2,477 operations), 6-25-04 (2,284 operations) and 3-6-04 (1,647 operations), have a recorded FAA/APO/ASPM efficiency. For these three dates the simulated and observed efficiencies appear sufficiently similar; therefore the simulation is accepted as a reasonable approximation of DFW operations. Based on the measured efficiency, there is sufficient room for growth to handle more flights at DFW than the forecast 2808 operations per day in year 2010.

### Table 2. DFW Performance Statistics

| Scenario | Wind conditions | PT status | Operations per day | VS Efficiency | FAA/APO Efficiency |
|----------|-----------------|-----------|--------------------|--------------|--------------------|
| 1 | South flow | Without PT | 2477 | 43.2 | 42.3 |
| 2 | North flow | Without PT | 2477 | 38.6 | - |
| 3 | South flow | With PT | 2477 | 44.6 | - |
| 4 | North flow | With PT | 2477 | 37.4 | - |
| 5 | South flow | Without PT | 2808 | 44.2 | - |
| 6 | North flow | Without PT | 2808 | 43.4 | - |
| 7 | South flow | With PT | 2808 | 50.8 | - |
| 8 | North flow | With PT | 2808 | 44.1 | - |
| 9 | South flow | Without PT | 2284 | 47.3 | 43.8 |
| 10 | North flow | Without PT | 2284 | 35.7 | - |
| 11 | South flow | With PT | 2284 | 45.1 | - |
| 12 | North flow | With PT | 2284 | 47.2 | - |
| 13 | South flow | Without PT | 1647 | 31.6 | 31.0 |
| 14 | North flow | Without PT | 1647 | 25.7 | - |
| 15 | South flow | With PT | 1647 | 25.7 | - |
| 16 | North flow | With PT | 1647 | 25.9 | - |

## Critical Evaluation of Airfield Geometry

A detailed evaluation of the DFW runway and taxiway geometry was performed to identify problem areas that may require further study or analysis to develop operating procedures and guidelines. The principal rationale for introducing the PT is to reduce runway incursions, improve safety, and significantly reduce delay to airlines and passengers. The PT as planned, designed, and constructed, is expected to improve operating efficiency and increase arrival and departure capacity at DFW. The analysis is performed for the four quadrants of the airport after the PT is in place and in operation. The planned path from arrival runway to terminals, general aviation and cargo aprons are compared with the animated operations created from the VS simulation.

The taxiway geometry and the links to the proposed PT required a critical evaluation from the Ground Controller's and Local Controller's point of view. Evaluation of the flow and movement of aircraft on the PT configuration for the South Flow and the North Flow is performed separately to identify the areas that require further study to eliminate runway incursions and collision avoidance. Pilots should have situational awareness and use extreme caution and allow sufficient distance between aircraft while traveling on the PT. The simulation used the minimum separation between aircraft built-in the VS program based on the FAA criteria and the speed of travel on the PT was set at 15 mph. The simulation and the animation showed that during South Flow, a large number of aircraft were using the SE quadrant of the PT; a similar situation arises on the NE quadrant PT during the North Flow operations that require careful evaluation and the development of a standard taxiway operational guidelines, procedures, and control.

### NE Quadrant Analysis

Figure 3 shows the movement of aircraft traveling to terminals A, C, E and general aviation area after landing on runways 35C, 35R and 31R during North Flow operations at DFW. The figure also shows a B767-300 departing from runway 35L overflying a B737-300 on the PT. This section of PT feeds five terminals and a large number of aircraft are moving on the PT during peak periods. There are aircraft exiting from the FedEx and east cargo area that take taxiway P, travel south to join the departure queue on the east side of 35C for take off to the north. The intersection of taxiways P, Q, R and N is a choke point where both arriving and departing aircraft meet during North Flow operations. Therefore, a detailed evaluation of this choke point needs to be done to properly regulate the movement of aircraft and develop detailed standard taxiway procedures and guidelines for pilots.

### SE Quadrant Analysis

Figure 4 shows the operation of the SE quadrant PT during South Flow configuration. Aircraft arriving on 17L and 17C use Taxiway P to taxi on the PT to reach terminal buildings on the east and west side. This is the busiest section of the PT that receives aircraft from runways 17C and 17L during South Flow operations. The aircraft traveling south from Taxiway P will reach the choke point at the intersection of PT and Taxiway M. The Ground Controller will direct traffic at this choke point based on the standard taxiway procedures and guidelines. Each aircraft will be given clearance to cross the choke point and allowed to taxi to the next hold point and wait for instruction from Ground Controllers to proceed further.

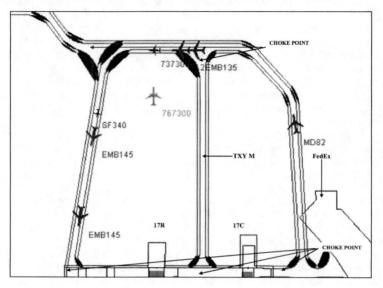

**Figure 3. NE Quadrant PT**

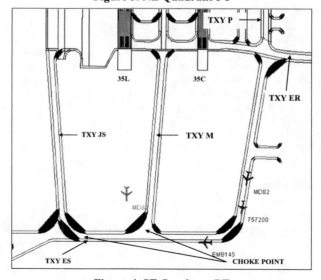

**Figure 4. SE Quadrant PT**

### SW Quadrant Analysis

During the South Flow operations (see Figure 5), the aircraft landing on 13R and 18R take the high speed exit and travel on the PT to reach terminals B and D on the west side. Aircraft traveling to terminals A, C and E use the Taxiway A bridge on the south side and turn left on Taxiway K to head north.

During the North Flow operations, the aircraft departing from the UPS apron, and West air cargo aprons travel south on Taxiway C to join the departure queue for runway 36L from the west, which requires crossing the arrival runway. In Figure 5, a B747-400 is heading south to runway 36L, aircraft, SF340, is entering the departure queue on runway 31L, another aircraft SF340 is traveling north on taxiway C to runway 31L. This portion of Taxiway C is designated in VS as a Dynamic Single Direction (DSD) path allowing one aircraft only in the link from PT entrance to Taxiway WM.

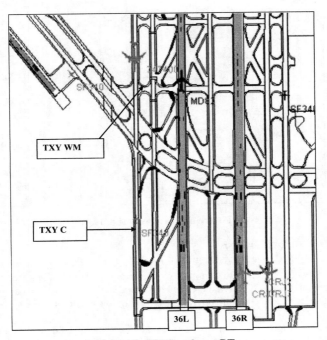

**Figure 5. SW Quadrant PT**

### West side Taxiway C Analysis

In Figure 6, aircraft B777-200 landed on 36L, and exited on the high speed exit, but it had to come to a complete stop to allow the B737-200 aircraft heading to the west cargo apron area through the intersection. In the VS simulation, this section of Taxiway C has been designated as a DSD that permits only one aircraft on the specified link on Taxiway C from the UPS apron to Taxiway WK. There are departing flights from UPS that travel to the west side departure queue on Runway 36L for takeoff during North Flow operations. This part of Taxiway C requires a detailed evaluation and standard taxiway procedures and guidelines developed to control the movement of aircraft on this section of Taxiway C.

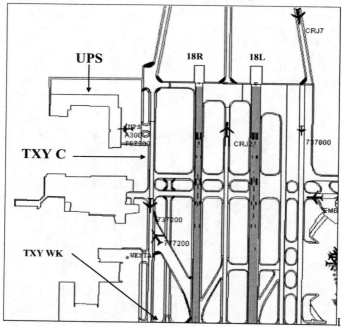

**Figure 6. Taxiway C Analysis**

### East side Taxiway P Analysis

Figure 7, shows a B777-300 is exiting on the high speed exit from runway 17C to taxiway P traveling to terminal A. The requirement with the introduction of a PT is that heavy aircraft should take the high speed exit before the hangar on the east side of taxiway P and continue on Taxiway P south to travel on the PT to the terminal gates.

This portion of the taxiway requires the development of detailed procedures and guidelines for arriving pilots to watch for aircraft exiting from the hangar on the east side. If the cargo aircraft is heavy (Group V) arriving on 17C, it has to take the high speed exit to taxiway P, then head north to east cargo apron, or if the aircraft is large, it has to take the high speed exit to taxiway M to travel south on the SE PTs . The heavy aircraft will make a left turn from the high speed exit, and go north on Taxiway P to the cargo apron on the NE end freight area. The large aircraft will use the PT to taxi on Taxiway P to head north to the cargo apron on the NE quadrant of

DFW. Therefore, it is suggested that all cargo aircraft should be directed to land on 17L and after exiting from the runway they will travel north on taxiway Q to reach the east freight and FedEx aprons, thus avoiding conflict on taxiway P with the arriving heavy aircraft destined to terminal buildings on the west side of runway 17C.

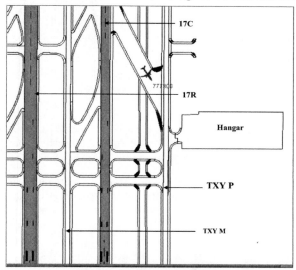

**Figure 7. Eastside Taxiway P**

### *South Flow arrivals*

During South Flow operations, the NE quadrant PT and the NW quadrant PT do not carry any aircraft taxiing to the terminals. The only aircraft taxiing in the NE quadrant of the PT are turboprops heading to runway 13L for takeoff as shown in Figure 8, where an MD82 is overflying a SF340 on the PT taxiing to runway 13L. During South Flow configuration, there were 408 aircraft arriving on runway 17C and 74 aircraft assigned to departure runway 13L for departure. There are no aircraft on the NW quadrant PT traveling to any terminals or cargo aprons. This is true for the SW quadrant PT and SE quadrant PT during the North Flow operations. Therefore, the arrival aircraft flying over the PT do not encounter many aircraft as opposed to the departure situation where the departing aircraft has to overfly all types of aircraft traveling on the PT towards the terminals.

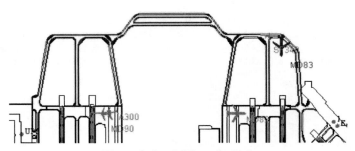

**Figure 8. South Flow Arrivals**

### North Flow arrivals

North Flow arrivals are shown in Figure 9 over South PT, where one MD 82 is overflying the SW quadrant PT, and another MD 82 is overflying the SE quadrant PT. There were 457 aircraft arriving on runway 36L and 72 aircraft assigned to runway 31L for departure.

The FAA had simulated the operation of the arrival flights over the PT with video of several aircrafts taxiing on the PT and an aircraft overflying them. This research found that the arriving flights do not encounter as many aircraft as perceived by the FAA.

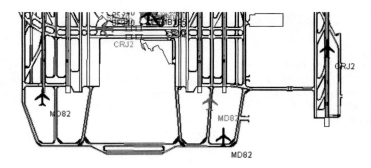

**Figure 9. North Flow Arrivals**

### Discussion of AOSC decision on PTs

The FAA/AOSC decision on PT design and construction was approved on June 8, 2005 [AOSC, 2005]. In this document, the AOSC team restricted the height of aircraft on the PT to be at 65 ft at a distance of 2,650 ft. (40:1 slope) for all weather departure of Group V aircraft during South Flow on Runway 17R. An aircraft with a tail height greater than 65 ft will not be allowed to use the PT without specific instruction from the ATC. In the VS simulation, all types of aircraft were allowed to use the PT in all four quadrants. An analysis of the flight tracks over 17R/35L and 18L/36R showed that the departing aircraft reaches high altitude by the time they cross the PT centerline at a distance of 2,650 ft on the North and South side of DFW.

## Conclusions

The PT operations at DFW, once completed in the future should reduce runway incursions; however, the taxi in time may increase during peak hours of operation as aircraft have to be directed by the Ground Controllers to join the queue to reach the terminal building. The choke points, which are identified in this paper, and several may cause a slow down in the movement of aircraft during arrivals and departures. Based on the operational analysis, a careful study of planned cargo facilities must be undertaken to identify their impact on PT operations. This research should pave the way for simulation of real flight data using VS for ORD, DTW, LAX and STL toward implementing a PT system to improve runway safety. The impact of PT operations on individual airlines and the cargo carriers require further evaluation.

## References

Airbus S A S (2005) "Traffic Forecast- Global Market Forecast 2004-2023", Airbus Industries, Paris, France.

Airport Obstructions Standards Committee (AOSC) (2005) "Dallas Fort Worth (DFW) End-Around Taxiway System." AOSC, Decision Document # 06, FAA, Approved, June 8, 2005

Associated Construction Publications, Inc. (2006) "e-regional Contractor", Irving, Texas http://www.acppubs.com/index.asp?layout=noCClamp&articleID=CA638192 5. Web site accessed on 11-24-06

Boeing Commercial Airplanes (2005) "Current Market Outlook 2005" World demand for commercial airplanes, Market Analysis, Current Market Outlook, Boeing, P.O. Box 3707, Seattle, WA 98124-2207.

Buondonno, Karen and Kimberlea, Price.(2003). "Dallas/Fort Worth International Airport Perimeter Taxiway Demonstration." U.S. Department of Transportation, FAA, Washington D.C. 20591 Doc No DOT/FAA/CT-TN03/19

Davis, William (2002) "Perimeter Taxiways and Improved Surface Safety" Director, Runway Safety Program, Office of Runway Safety, FAA Washington D C. 20591

Davis, Williams (2003) "Perimeter Taxiways" FAA Office of Runway safety, FAA, Washington D C, 20591.

DFW Airport (1996) "Runway Use Plan" DFW Airport, P.O. Drawer 619428, TX 75261-9428

Erway, Paul S. (2003) " Runway Safety Program" Runway Safety Program Manager, Southwest region, FAA, Fort Worth, TX, 76137

Federal Aviation Administration (2005) "FAA Aerospace Forecasts- Fiscal Years 2005-2016", FAA, 800, Independence Avenue SW, Washington, D.C. 20591.

Federal Aviation Administration (2004a) "Flight Plan 2005-2009" FAA, 800, Independence Avenue SW, Washington, D.C. 20591.

Federal Aviation Administration (2004b) "Terminal Area Forecasts

FAA Office of Runway Safety (2003) "FAA Runway Safety Report: Runway incursion trends at Towered Airports in the United States, FY 1999-2002." FAA, 800, Independence Avenue SW, Washington, D.C. 20591

Leigh Fisher Associates (1996) "Assessment of Runway Crossing delays and Runway Reconstruction Alternatives Dallas/Fort Worth International Airport" Working Paper, Prepared for DFW International Airport Board, DFW, Texas, 75261

North Central Texas Council of Governments (NCTCOG) (2003) "North Central Texas 2030 Demographic Forecast" NCTCOG, Arlington, Texas, 76005

Wine, Carlton (2005) "Presentation of System Airport Efficiency Rate (SAER) and Terminal Arrival Efficiency Rate (TAER)" Presented to Customer Satisfaction Metrics Work Group, FAA, 800, Independence Avenue SW, Washington, D.C. 20591

# Using Fuzzy Multi-Criteria Decision Making Approach to Evaluate Airport Facilities Maintenance Policy

Yu-Hern Chang,[1] and Yi-Cheng Pan[2]

[1]Professor, Department of Transportation and Communication Management Science, National Cheng Kung University, 1 Ta-Hsueh Road, Tainan, Taiwan, R.O.C.; PH (886) 6275-7575#53224911-1234; FAX (886) 6275-3882; email: Yhchang@mail.ncku.edu.tw

[2]PhD student, Department of Transportation and Communication Management Science, National Cheng Kung University, 1 Ta-Hsueh Road, Tainan, Taiwan, R.O.C.; PH (618) 303-4764; email: panyicheng@gmail.com

## Abstract

Globalization and technology development have led to rapid world-wide growth, which has in turn significantly enhanced competition among airports around the world. Competitive threat of privatized airports and second-tier airports create pressures on airport's operation. In order to increase revenue and decrease expenditure, more and more airport managers are searching for an optimal and customized maintenance management concept policy. There is a desire to improve the airport safety and quality and increase mean-time-between-failure (MTBF) of facilities to achieve an optimal customized maintenance management. The research reported in this paper, adapted a framework for maintenance concept developed for industrial management. In addition, the paper offers some guidelines to develop maintenance management concept policy. Furthermore, the paper identifies the effectiveness of using a fuzzy multi-criteria decision making (FMCDM) approach to evaluate maintenance management concept policy based on five basic maintenance concept policies. The approach discussed in the paper will help airports develop appropriate maintenance management and increase sustainable competitive advantage.

## Introduction

The undeniable global competition, characterized by both a technology push and a market pull, and the rapidly evolving technology and increased passenger requirements has created numerous challenges for airport management. One of these challenges concerns the airport facilities maintenance management. The airport maintenance decision-making process currently suffers from a large fragmentation between airside and landside operations and the isolated manner in which the airport maintenance performance measures are assessed and handled (Zografos & Madas 2006). High passenger growth and utility rate put airport facilities under higher stress. In this situation, once the equipment is out of order and control, it will affect the flight safety and airport managers will incur a lot of cost and have to spend time to deal with the failure. In the meantime, it will cause a chain reaction resulting in reduction of safety and increased passenger and airlines complaints against airport. In order to deal with this problem, an airport's strategic investments in airport facilities should not only consider cost and capacity, but also consider technology trends, flexibility, and other factors (Jarach 2005).

An airport contains a large number of technical systems on airside and landside, which act interdependently to achieve the desired safety and quality of aviation services (Min 1994). Effective maintenance plays a critical role in the achievement of safety and quality objectives. Proper maintenance not only ensures flight safety; it also improves the overall performance of the airport. However, maintenance also adds significantly to the total cost, and this often forms the basis of analyzing the performance of the maintenance department. The decision on the required maintenance management concept and a thorough and easily accessible technical knowledge are crucial here. In addition, the increasingly congested major hub airports face increasing competition from the privatized airports and second-tier airports (Barnard 1998). For these reasons, all major airports worldwide are searching for an optimal and customized maintenance management concept and policy. In order to help the airports solve these problems, this study adapted a framework for airport maintenance management evaluation proposed by Waeyenbergh and Pintelon (2002). The adapted framework also uses a fuzzy multi-criteria decision making (FMCDM) approach to evaluate maintenance management based on five basic maintenance concept policies.

## Need for a Customized Airport Maintenance Concept Policy

In today's aviation industry, everything has to be fast. In order to be able to meet the passenger's demands, aviation technology development has to be fast, flight capacity has to be great, flight time has to be quick, and flight slot has to be short (Gordon 1986). In addition, maintenance has to be performed rapidly and efficiently, because one of the key factors responsible for large losses of performance and profit is downtime. To achieve high availability, good system reliability, maintainability and safety, an effective maintenance management concept policy is essential for airport management. To ensure that the airport facilities availability is stable and increasing, it is very important to keep up to date with the development of a maintenance management. This involves an abstraction of meanings from reality which is understandable for others and which explains, guides and controls how the maintenance process happens or works. Proper maintenance management helps to keep the maintenance life cycle cost down, ensures proper and safe operations and streamlines internal logistics.

Each airport obviously has individual interests in the facility design. However, in order to develop a functionally efficient maintenance that met airport's short and long range operational goals at a reasonable cost several key factors have to be taken into account by the design team (Rivera & Sheahan 2005). However, to make rational and justifiable tactical decisions concerning maintenance, one needs to have a clear idea of the advantages and disadvantages of each maintenance management concept policy. In addition, a supporting maintenance management concept policy is required. Developing and implementing a maintenance management concept policy is a difficult process that suffers from the lack of a systematic and consistent methodology or a framework. This paper discusses the effectiveness of using an FMCDM approach to evaluate maintenance management concept policy, which can in turn help airport maintenance managers in developing appropriate maintenance management schemes.

## Overview of Airport Facilities

Airport facilities vary from one airport to another and in general can be categorized as airside facilities, landside facilities and other on-airport facilities. Generally, airside involves facilities for aircraft landing, takeoff, and taxi functions. Such facilities include main runway, secondary runway, high/medium intensity runway lights, instrument landing system (ILS) [consisting of glide slope, localizer, and a short approach light system with sequenced flashing lights (SALSF)], precision approach path indicator (PAPI) lights, runway end identifier lights, visual approach slope indicator (VASI) lights, taxiway system and others. The nature and scale of facilities depend on the specific characteristic of airports (King County International Airport 2007).

Landside facilities are varied and different based on airport's environment and scope of operation. Commonly, landside facilities include aircraft parking aprons, passenger terminal facilities, industrial aerospace facilities, air cargo facilities, storage hangars, maintenance hangars, air traffic control tower facilities, and automobile access and parking.

Other on-airport facilities are ancillary facilities such as aircraft rescue and fire fighting (ARFF) facilities, fuel storage facilities, airport maintenance facilities, Airport Administrative Offices, and others.

## Framework for Airport Maintenance Concept Policy Development

To develop an appropriate maintenance management concept policy, maintenance must be considered holistically. The policy should consider factors that technically describe each system to be maintained, factors that describe the interrelationship among different systems and factors that describe the general organizational structure. If some of the necessary aspects are not considered (e.g. due to careless analysis, loss of data or lack of knowledge), the maintenance management will never reach its full potential. The general organizational structure in maintenance interventions makes the importance of an appropriate maintenance management concept policy quite clear. Because of the high direct and indirect cost involved (for in-house as well as for outsourced maintenance) and because of the operational impact maintenance may have on the equipment's performance, maintenance management concept policy development should be done in a structured way. Moreover, the maintenance management should be customized, i.e., it should consider all relevant factors of the situation on-hand. As such, it will be really tailored to the needs of the airport in question. This means that the maintenance management concept policy will be unique for each airport.

The underlying structure for developing such a management policy may however be very comparable. Another important remark is that since aviation industrial systems evolve rapidly (for example, consider the CAS/ATM technological innovation), the maintenance management will also have to be reviewed periodically in order to take into account the changing systems and the changing environment. This calls not only for a structured, but also for a flexible maintenance management concept policy, allowing feedback and improvement.

Taking into account all these requirements, it was decided to develop a customized maintenance concept policy using a framework adapted from the one developed at the Centre for Industrial Management (CIB). The origin of this framework has been described in Waeyenbergh and Pintelon (2002). It has different modules. The first module is the start-up module. In this module, identification of the objectives and resources in airport will take place. In the second module, the technical analysis takes place. The Most Important Systems (MISs) and the Most Critical Components (MCCs) in airport facilities will be identified. In the third module the appropriate maintenance policy will be chosen and fine tuned. The fourth module is the module of maintenance policy implementation and evaluation. Depending on the output of this fourth module, the fifth module, the module of continuous improvement, will act on the first three modules. Thus, seven steps can be distinguished in the framework (see Table 1 below).

**Table 1. 7-Step Framework**

| | |
|---|---|
| Step 1. Identification of the all facilities in airport | Module 1 |
| Step 2. Selection of the Most Important Systems in facilities | Module 2 |
| Step 3. Identification of the Most Critical Components | |
| Step 4. Maintenance policy selection | Module 3 |
| Step 5. Optimization of the maintenance policy parameters | |
| Step 6. Implementation and evaluation | Module 4 |
| Step 7. Feedback and improvement | Module 5 |

## Maintenance Management Concept Policy

Even though there are various maintenance concepts and policies discussed in past literature, some of which are proprietary and used by consultants, more and more airports are searching for their own customized maintenance management concept policy (Waeyenbergh and Pintelon 2002). The concepts described in literature are often very time-consuming to implement or only valid for a special class of equipment or a specific industry. Most of the concepts from literature, however, offer interesting and useful ideas. The framework proposed in this paper uses ideas of five maintenance concept policies. In this light, a few of the most important maintenance concepts and policies are not discussed here. The cost of operations and maintenance can make or break a business, especially with today's increasing demand on productivity, availability, quality, safety and environment, and the decreasing profit margins. In maintenance, there are two basic interventions: Corrective Maintenance and Preventive Maintenance (Lynch 1996). According to the way these two basic interventions are applied, five basic maintenance concept policies can be distinguished: Failure Based Maintenance, Design-Out Maintenance, Use Based Maintenance, Condition Based Maintenance and Detection Based Maintenance (Waeyenbergh and Pintelon 2002).

In Failure Based Maintenance (FBM), maintenance concept is carried out only after a breakdown. Design out maintenance (DOM) is to improve the design in order to make maintenance easier or eliminate the need for maintenance. Use Based Maintenance (UBM) is carried out after a specified amount of time (for example, after 3 months, 3000 working hours, etc.) such as car maintenance. Condition Based Maintenance (CBM) defines as a consequence of a follow-up of the technical installation with high-tech monitoring techniques. However maintenance as a consequence of the detection by operators will be called DBM (Detective Based Maintenance). If no maintenance can be carried out when a certain irregularity (detectable with the human senses) is observed by the operator/technician, the more expensive option of CBM has to be considered.

## Constructing a Multi-criteria Decision-making Model

### Model

Before making a multi-criteria decision, one must consider all impact on multiple dimensions. Each dimension has its own multiple evaluation criteria, which form a hierarchical multi-tier problem structure. Many scholars use the Analytic Hierarchy Process (AHP) method to deal with strategy and policy selection problems. A fuzzy notion was introduced into the AHP method. Instead of asking the survey respondents to select a specific utility score, this notion allows a range of utility scores. Such method was used to study the marketing strategies of the information service industry (Tang and Tzeng 1999) and to analyze the strategy choices of IC and communication companies in Taiwan (Yu et al. 2005). This research uses the fuzzy AHP method to select customized maintenance concept policies.

Different companies were surveyed to understand why one company chose one maintenance concept policy over others. In addition, a brainstorming exercise with experts in the maintenance arena helped establish what is expected from a good maintenance concept policy. Three top-level dimensions that a maintenance concept policy must evaluate were breakdown consequence, cost, and importance. Under Breakdown consequence dimension, three evaluation criteria used were: (1) Breakdown consequences for safety, (2) Breakdown consequences for the airport operation, (3) Possibility for secondary damage (e.g. on other machines).Under cost dimension, three evaluation criteria used were: (1) Loss of production, (2) Repair cost (man-hours), (3) repair cost (material).Under importance dimension, we four evaluation criteria used were: (1) Ease of failure detection, (2) Is the system a bottleneck (3) Complexity of the system, (4) Redundancy back up system.

A multi-criteria decision-making model with three first-tier dimensions and ten second-tier evaluation criteria is shown in Fig. 1.

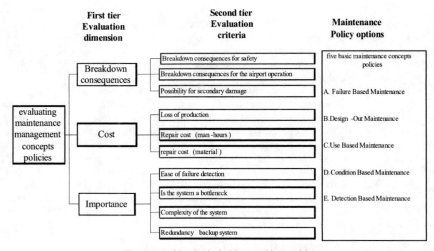

Fig. 1. A multi-criteria decision-making model

**Figure 1.  A Multi-Criteria Decision-Making Model**

### *Solution of Multi-criteria Decision-Making Model*

This paper used the FMCDM model to evaluate five maintenance concept policies to determine how well each of them met the 10 criteria. Experts in the aviation and airport maintenance arena were convened to determine what is expected from good maintenance concept policy parameters; they are the most knowledgeable in maintenance management. In the weighting factor design, we asked experts about his/her perception of the ratios of relative importance between pairs of the ten multi-criteria. A set of ten weighting factors associated with the ten multi-criteria can therefore be derived as follow:

Let $W$ (1) be the vector of weighting factors of subject n

$$W = (w_1, w_2, .... w_{10}) \tag{1}$$

For the five maintenance concept policies of interest, we wish to derive, from each expert, a utility score that represent the level of evaluation for each of the ten multi-criteria. We asked the "breakdown consequences" subjects/experts to select one out of five linguistic variables from: "very highly serious", "highly serious", "fairly serious", "low serious", and "very low serious" as a way to measure the level of evaluation for a criterion. In the cost and importance dimensions, subjects were to select one out of five linguistic variables: "very high", "high ", "fairly ", "low ", and "very low " as a way to measure the level of evaluation for a criterion. Because the perception or the interpretation of these linguistic variables is likely to be different for each subject, this study used the notion of triangular fuzzy numbers (TFN) to represent each survey subject's perception of the linguistic variables. We asked each survey subject to assign a fuzzy range of utility scores between 0 and 100 with lower estimate LE, medium estimate ME, and Upper estimate UE. We adopted the Center of Area (COA) method to convert the fuzzy range of utility scores to a non-fuzzy utility score (NFUS). We used the following method: The utility scores constitute a 10 x 5 matrix $\overset{\text{r}}{S}$ (2), where $S_j^k$ is the utility score of the jth evaluation criterion for policy k.

$$\overset{\text{r}}{S} = \begin{bmatrix} S_1^A & S_1^B & S_1^C & S_1^D & S_1^E \\ S_2^A & S_2^B & S_2^C & S_2^D & S_2^E \\ ... & ... & ... & ... & ... \\ ... & ... & ... & ... & ... \\ S_{10}^A & S_{10}^B & S_{10}^C & S_{10}^D & S_{10}^E \end{bmatrix} \tag{2}$$

The utility score matrix $\overset{\text{L}}{S}$ (2) will be multiplied by the vector of weighting factors of $W$ (1), to get $\overset{\text{L}}{U}$ (3), where $u^k$ is the utility score weighted by the weighting factors for policy k.

$$\overset{\text{L}}{U} = \overset{\text{L}}{W} \cdot \overset{\text{L}}{S} = (u^A, u^B, u^C, u^D, u^E) \tag{3}$$

Where $u^A$ is the utility score of policy k .The maintenance policy with the highest utility score is the most favored maintenance policy by that subject.

**Results of the FMCMD Model**

*Weighting factors*

All of the expert survey respondents were considered as one group. The weighting factors for the three first-tier goals are then: (1) Breakdown consequences (0.55); (2) cost (0.24); and (3) Importance (0.21). The importance of satisfying consumers' needs is greater than the combined importance of the other two. The priorities of the evaluation criteria used to measure maintenance concept policy parameters are: (1) Breakdown consequences for safety (0.49); (2) Breakdown consequences for the airport operation (0.14); and (3) Possibility for secondary damage (e.g. on other machines) (0.37). The weighting factors of the ten evaluation criteria for the group of all survey respondents are shown in Table 2.

*Utility Scores of Five Maintenance Management Concepts Policies*

The vector of weighting factors by all survey respondents is denoted as $\overset{\text{L}}{W}$ (4):

$$\overset{\text{L}}{W} = (0.2695, 0.077, 0.2035, 0.984, 0.0672, 0.0744, 0.0399, 0.0672, 0.0483, 0.0546) \tag{4}$$

**Table 2. Weighting Factors by All Survey Respondents**

| Dimension | Evaluation criteria | | |
|---|---|---|---|
| | Weighting factors of dimension | Weighting factors of evaluation criteria within a dimension | Weighting factors of evaluation criteria across dimensions |
| Breakdown consequences dimension | 0.55 | | |
| (1)Breakdown consequences for safety | | 0.49(1) | 0.2695(1) |
| (2)Breakdown consequences for the operation | | 0.14(3) | 0.077(4) |
| (3)Possibility for secondary damage (e.g. on other facilities or systems). | | 0.37(2) | 0.2035(2) |
| Cost dimension | 0.24 | | |
| (1) Loss of production | | 0.41(1) | 0.0984(3) |
| (2) Repair cost (man-hours) | | 0.28(3) | 0.0672(7) |
| (3) repair cost (material) | | 0.31(2) | 0.0744(5) |
| Importance dimension | 0.21 | | |
| (1) Ease of failure detection | | 0.19(4) | 0.0399(10) |
| (2) Is the system a bottleneck? | | 0.32(1) | 0.0672(6) |
| (3) Complexity of the system | | 0.23(3) | 0.0483(9) |
| (4) Redundancy. | | 0.26(2) | 0.0546(8) |

The product of $\overset{\perp}{W}$ (4) and the score matrix $\overset{\perp}{S}$ (2) of calculates the utility scores of the five maintenance management concepts policies. The maintenance policy with the highest utility score is the most favored maintenance policy by that component.

Once the decision on the type of maintenance management policy has been made, the parameters of this policy must be optimized by continuing improvement. In the past, many companies limited maintenance performance reporting to a minimum budget reporting. The main reason for this is that maintenance performance reporting is difficult.

There is a time-lag effect, which makes it difficult to specify the amount and intensity of the service (and the corresponding required amount of money) needed for assuring proper plant performance. Still another aspect which makes it difficult to measure maintenance output is the fact that maintenance is closely related to other activities. Both the merits and shortcomings of the service rendered are not immediately apparent. However, in order to anticipate problems and opportunities to make the necessary adjustments, sound performance reporting is indispensable. Moreover, efficient performance reporting systems support continuous improvement.

Performance reporting systems are based on performance numbers or indicators. They measure the performance concerning a specific aspect of operations or planning in a given period. These resulting numbers are then placed on a scale in order to evaluate that performance. This paper developed a framework for airport maintenance management concept policy evaluation to help airport continuous improvement to achieve near zero downtime and closed-loop life cycle design.

## Conclusions

More and more airports are looking for a customized maintenance concept policy. The framework presented in this paper sets out guidelines for establishing an airport maintenance concept policy. The proposed framework is flexible to allow enough customization space to create an airport-specific concept, which could incorporate new knowledge and information from new information and communication technology.

This paper has provided the conceptual framework for using the fuzzy multi-criteria decision making approach to assist the airport managers to select the maintenance management concept policy. The inherent dilemma can be broken if airport manages few important systems and components as their maintenance policy to achieve the overall goal. The goal of any maintenance is to increase throughput and at the same time reduce inventory and operating expense. As a prerequisite to ensuring profitability, the airport manager must be able to quickly identify and remove the constraint(s) and ensure that they can continue to meet airport's changing environment accurately. However, using fuzzy multi-criteria decision making approach to evaluate maintenance management concept policy parameters should be adopted with care due to its intensive training requirements and radical approach that requires experimental learning.

In future development of the framework, more evaluation parameters (more specific in categorized airport facilities) should be added such as airside evaluation parameters and landside evaluation parameters, and the content of each module should be carefully elaborated upon to components. In this way, a modular framework can be built. This modular approach would allow airports to incorporate ideas (i.e. a selection of evaluation parameter modules) in order to make a customized maintenance concept policy in computerized information and knowledge system.

## Acknowledgments

Authors acknowledge maintenance managers of CHINA AIRLINES, FAT EASTERN AIR, TRANSASIA AIRWAYS, UNI AIR, AIR ASIA, AIDC, EGAT, and International CKS Airport.

## Reference

Barnard, B. (1998). "Europe leads global airport race." *Europe*, 373, 12-14.

Gordon, J. K. (1986). "Computerization Promotes Integrated Airport Management." *Aviation Week & Space Technology*, 125 (18), 138-141.

Jarach, D. (2005). *Airport Marketing: Strategies to cope with the New Millennium Environment,* Ashgate, Italy.

King County International Airport, (2007). "Inventory of Existing Airport Facilities" http://www.metrokc.gov/airport/plan/ (January, 12, 2007)

Lynch, M. (1996). "Preventive versus corrective maintenance." *Modern Machine Shop*, 69 (6), 144-146.

Min, H. (1994). "Location planning of airport facilities using the analytic hierarchy process." *Logistics & Transportation Review,* 79-94.

Rivera, T. and Sheahan, R. E. (2005). "More than just a maintenance facility: The DFW airport maintenance and storage facility." *International Conference on Automated People Movers,* May.1.

Tang, M. T. and Tzeng, G. H. (1999). "A hierarchy fuzzy MCDM method for studying electronic marketing strategies in the information service industry.", *Journal of International Information Management,* 8 (1), 1–22.

Waeyenbergh, G., and Pintelon, L., (2002). "A framework for maintenance concept development." *International Journal of Production Economics* 77, 299 – 313.

Yu, H. C., Lee, Z. Y., and Chang, S. C. (2005). "Using a fuzzy multi-criteria decision making approach to evaluate alternative licensing mechanisms." *Information & Management,* 42(4), 517-531.

Zografos, K.G., and Madas, M.A. (2006). "Development and demonstration of an integrated decision support system for airport performance analysis." *Transportation Research: Part C,* 14 (1), 1-17.

# Building Sound Emergency Management into Airports

James Fielding Smith P.E., PhD, M.ASCE,[1] Sandra S. Waggoner BA, EMT-P, EMSI,[2] and Gwendolyn Hall, PhD[3]

[1]Professor of Environmental, Emergency, and Disaster Management, American Public University System, and partner, JSW Associates, 385 Sam Reed Rd NW, Floyd, VA 24091-3551; PH (540) 763-3068; FAX (540) 763-3268; email: jim@jswassociates.cc
[2]Partner, JSW Associates, email: sandy@jswassociates.cc
[3]Dean, School of Public Safety, National and Homeland Security, American Public University System, 111 W. Congress St., Charles Town, WV 25404; PH (304) 724-2813 ; FAX (304) 724-2858 ; email: ghall@apus.edu

## Abstract

Airports are essential and irresistible assets in major disaster responses. Traditional and newly emergent roles—e.g., command posts, shelters, temporary hospitals, --alternative communication hubs—were filled by airports after Hurricane Katrina and for 9/11 flight diversions with essentially no warning. The basic thesis of this paper is that sound emergency management measures should be built into airport preparedness functions to not only maximize the use of airports during major disaster responses but also to preserve airport operations during the disaster and to facilitate resiliency afterwards. Qualitative analysis applied to historical case studies at New Orleans and Gander and preparedness measures at Memphis and Düsseldorf gives results bearing on necessity to enhance capabilities of airports and methods to fund and sustain such new assets. National standards and new sources of funding to build sound emergency management into airports are critical to the development of resiliency through timely management of catastrophes.

## Introduction

Airports are critically consequential in the modern world because of the role they play as conduits for travel and commerce, as employers, as economic development engines, sources of civic pride, hubs for transportation operations, land use activities, dual use with military units, hosts to aviation-related industries, pillars of just-in-time logistics, and parts of integrated business and resort developments. Since September 11, 2001, and Hurricane Katrina, the world has become aware that airports are crucial to emergency response and recovery activities that extend far beyond incidents directly related to traditional activities of airports.

47

The events of 9/11 suddenly transformed every airport in the world into a junction of air transportation and emergency response. Few airports were ready with capabilities to support this new role. Hurricane Katrina demonstrated the need for revision of airport operational and design doctrine. Louis Armstrong New Orleans International Airport became the nerve center for command and control by housing various rescue operations and became the site for massive medical operations, a multimodal beach head, a base for command and logistics, as well as a press center and aid reception site. At the same time, it had to sustain commercial air traffic in addition to increased regular flight loads.

The events of 9/11 led to the adoption of federal template for emergency response with the development of the National Response Plan (NRP). The National Incident Management System (NIMS) and National Infrastructure Protection Plan (NIPP) set procedural goals and objectives to facilitate improved emergency management in disasters. For the details of NIMS and its philosophy see Homeland Security Presidential Directive 5 (White House 2002).

The goal of this study is to examine how roles for airports interface during the response and recovery phases of a disaster, placing special demands on an airport's spaces and communications which may compromise the airport's resiliency. The question is whether or not airports need or deserve special assistance to pay for the planning, preparedness, and mitigation to make these functions achievable.

The primary thesis of this study is that sound emergency management should be built into airports and their operations to better serve citizens, optimize asset utilization, and protect airports' return to normal operations. A major airport by its location, structures, security measures, and connections to community infrastructure is already well suited to fulfill many roles in emergencies. Airports can serve well as incident command posts, emergency operations centers, and multi-agency coordination center sites. The multimodal infrastructure facilities and physical security systems facilitate uses as beach heads for military aid, self-defense, and large-scale logistics support during operations. In addition, a major airport has the capacity to offer alternative communications and intelligence systems. The same systems that support normal airport operations could, if interoperable with first responder systems, support comprehensive operations throughout response and recovery.

This study relied on analysis and synthesis of information concerning five major case studies illustrating various aspects of disaster response or preparedness-- Louis Armstrong New Orleans International Airport (MSY), Gander International Airport (YQX), Los Angeles International Airport (LAX), London Heathrow Airport (LHR), and Memphis International Airport (MEM). All were sites of recent use for disaster management or have taken steps towards preparing for an enhanced role in incident management or both.

Qualitative analytical methods applied to information in the case studies allowed comparing and contrasting it to NIMS standards. The results are displayed in separate tables for each airport. Technical information, airport operational statistics, and budget information were not collected for all cases.

## Results

### *Louis Armstrong New Orleans International Airport (MSY)*

Table 1 gives the results for MSY.

| Airport | Major Regional Threat(s) | Actual Response Uses and Mitigation Measures | Future Greatest Regional Threat(s) | Planned Responses and Mitigation Measures Based on Airport Plans and Documents |
|---|---|---|---|---|
| Louis Armstrong New Orleans International Airport (MSY)<br><br>Owner: New Orleans Aviation Board<br>www.flymsy.com<br>State: Louisiana<br>Country: USA | Hurricane Katrina | Provide physical facilities<br>Medical services<br>Continued "normal" airport operations for passengers, freight, and flight support<br>Refugee services and processing<br>Beach head<br>Emergency response operations<br>Military operations<br>Own-force protection<br>Sheltering<br>Logistics<br>  Staging relief<br>  Receiving relief<br>Command and control<br>Communications<br>Fiscal/administrative services<br>Continuation of operations (COOP) | Hurricanes<br>Mississippi River floods<br>Pandemic | *Beach head<br>*Command and control<br>*Emergency response operations<br>*Logistics<br>  *Staging relief<br>  *Receiving relief<br>*Sheltering<br>*Medical services<br>*Own-force protection Intelligence<br>*Communications<br>*Refugee services and processing<br>*Fiscal/administrative services<br>*Provide physical facilities<br>*Continuation of operations (COOP)<br>*Continuation of business (COB)<br>*Continuation of government (COG)<br>†Military operations<br>*Continued "normal" airport operations |

**Table 1. Louis Armstrong New Orleans International Airport.**

\* indicates documented as actually budgeted or implemented.
† indicates imputed as budgeted or implemented from other sources.

After Hurricane Katrina in 2005, MSY served as an incident command post operations and logistics base, shelter reception center, and medical treatment area. It was the largest most affected airport in the region and underwent significant operational changes to serve in the Katrina response (House 2006:276). Prior to Katrina, 2005 had been MSY's peak traffic year; but after Katrina, passenger traffic dropped nearly 60 percent (July 2005 compared to December 2005). By January 2007, passenger traffic had returned to about 70 percent of pre-Katrina levels (NOAB 2006). This created significant financial stresses during disaster operations.

MSY was used as a key terminal for the **evacuation** of 13,000 victims with 129 airplanes. "Despite their overall success, airlift operations needed to feed into an overall management system. There were times when the military and the private carriers were duplicating efforts. Moreover, the coordination of all the parts was complex" (House 2006:123). "There was insufficient food, water, and sanitation" (House 2006:123). "…evacuees were being taken from a dehumanizing experience and placed into an equally dehumanizing environment at the New Orleans Airport" (House 2006:294).

MSY served as a major node for **communications** using various hosted mobile communications assets. MSY was not pre-wired for command and control functions and system disruptions hampered existing capabilities. DHS' National Command System (NCS) sent a satellite communications van to MSY and temporary cell phone service and satellite phones were also used (House 2006:177). MSY had no special standing communications capabilities to facilitate its use for non-airport actions. Crippled systems were overwhelmed. Structures blocked cellular signals even after installation of mobile towers (House 2006:294).

**Medical services** at MSY after Katrina have gotten the most attention, primarily because of their uniqueness, volume, and conflict. The entire medical operation at MSY treated perhaps as many as 8,000 patients eventually evacuated to other facilities (House 2006:294).

Much of the MSY space hosted medical units: Eight Disaster Medical Assistance Teams (DMATs), one thousand Commissioned Corps officers from the U.S. Public Health Service (PHS), an USAF Expeditionary Medical Support (EMEDS) from DOD, a Department of Health and Human Services (HHS) Federal Medical Shelter (FMS), local hospital staff, and volunteers, plus emergency medical services (EMS) and vehicles (House 2006:225). Units were typically established in tents in concourses. Triage was in the lobby and ticketing area. The whole lifespan of victims was represented: from the obstetrical unit to a temporary morgue (Woodruff, 2007), with medical diagnostic and treatment facilities and personnel for all stages in between (House 2006:271-276).

There were problems of coordination among all units using the airport. Working conditions for the medical units are described by two passages cited in the House Katrina report from official after action reports (AARs), one for the USAF EMEDS unit and one for OR-2 DMAT from Oregon (House 2006:276-277).

OR-2 DMAT's AAR (Miller 2005) documented a number of command-related issues. FEMA/NDMS implemented no form of an organized internal command and control structure at MSY, and there was no formalized unified command established between the many participating agencies until late in the response. Roles, responsibilities, and reporting structures were never clearly articulated. Liaisons among entities participating in relief efforts at the airport were never established and there was no initial interfacing at a management level resulting in friction among medical units. Inexperienced leaders were placed in an overwhelming environment that diminished effectiveness. Operational priorities were unclear. The OR-2 DMAT report further states, "FEMA/NDMS operations at the airport were extremely disorganized compared to parallel military and Forest Service operations…. Tensions between FEMA/NDMS and DMATs is an ongoing problem and continues to compromise the efficiency of operations due to a lack of trust between both parties" (Miller 2005).

Four days after Katrina hit land, problems in medical operations were judged bad enough that one major New Orleans hospital stopped sending patients. Local doctors were turned away as lacking recognizable credentialing required under FEMA, HHS, and DOD policies (House 2006:287). This is an example of the need to implement standardized credentials, which is a basic premise of multi-agency coordination.

The passages above from EMEDS and OR-2 DMAT AARs document in painful detail the lack of command and control at MSY. There was apparently no unified command (UC) implemented at any time. (House 2006:273) Major problems with coordination among and between medical units and other responders and MSY hosts complicated matters to the extent that a U.S. Forest Service command team was imported to supply command and control (House 2006:293).

One witness to the House Select Committee agreed there was tension and speculated that DMAT members are accustomed to commanding their own operations and are not used to taking orders from federal officials. He also said most of the FEMA NDMS commanders deployed were inexperienced, further contributing to the friction (House 2006:287, 293, footnote 264).

This study found no evidence that any airports had been involved in disaster planning for the operations of any of the medical units that responded to Katrina. The ultimate analysis was that "The medical operation at the New Orleans Airport was chaotic due to lack of planning, preparedness, and resources" (House 2006:287-294).

Within two days after Katrina's landfall, MSY began its roles in **logistics** for military **support to civil authorities** and **NGO relief agency** activities (House 2006:226). Various **law enforcement** activities used MSY and also protected the airport and its activities (House 2006:243). Federal Air Marshal Service personnel at the airport expanded their mission to include interim law enforcement activities as well as all necessary activities to operate the airport (House 2006:256, 264). One must also wonder at the Federal Aviation Authority (FAA) reaction to untrained, non-certified personnel handling ground traffic at the airport (Rabjohn 2007).

Despite its prominent role in these emergency response functions, MSY does not appear in any form in the conclusions to the House Select Committee Katrina report. The minority views appendix to that report emphasized basic command and control shortfalls in Katrina, but also did not mention MSY in their conclusions. Congressmen Melancon and Jefferson concluded that "FEMA management lacked situational awareness" and suffered from an "overwhelmed logistics system." They saw how "massive" communications inoperability "impaired response efforts, command and control, and situational awareness" (Melancon and Jefferson 2006).

Planning at MSY for the next major emergency emphasizes these considerations (NAOB 2006; Rodriguez 2006; Rodriguez 2007; and Woodruff 2007):

(1) Incident command and control requires cooperation by the airport in advance, during, and afterwards, but the implementation is almost totally out of the airport's control during incidents.

(2) Military integration into civilian response was less of a problem at MSY than anticipated.

(3) Successful command and control cannot begin mid-event; it has to be developed through joint planning and preparedness actions.

(4) The greatest vulnerability of any airport is physical hardness of the facility. In the case of MSY, the key areas probably could not have withstood more than a Cat 3 hurricane. Key facilities need to be built to survive the worst anticipated threat to the region.

(5) FAA certification and funding policies suitable for reconstruction and repair are not flexible enough to deal with upgrades much less expansion to support mitigation.

MSY asked the FAA to pay 100% instead of the usual 75% for the replacement ARFF, but the FAA declined. In the hurricane aftermath some projects were eligible for 100% but apparently the key factor was direct damage from Katrina. Therefore, FAA would pay for repairs but not improved capacity features. MSY hopes to find state funds for the local share and new state or federal funding to expand the ARFF, harden it for a Cat 5 hurricane, and provide space and pre-wired communications to handle area level organizational needs.

The Katrina-MSY story is still not complete in February 2007. Major Senate subcommittee hearings began on January 29, 2007, with an outcome that will not be known for months.

### Gander International Airport (YQX)

Table 2 presents the results for YQX. On September 11, 2001, all air traffic over and approaching North America was grounded to the nearest airport. Transoceanic flights were diverted before entering U.S. airspace to unaccustomed airports such as Gander and other Canadian airports (Fenton 2002; Vey 2007; Chang 2003). However, no reference to this appears in the 9/11 Commission report (9/11 Commission 2004).

**Table 2. Gander International Airport.**

| Airport | Major Regional Threat(s) | Actual Response Uses and Mitigation Measures | Future Greatest Regional Threat(s) | Planned Responses and Mitigation Measures Based on Airport Plans and Documents |
|---|---|---|---|---|
| Gander International Airport (YQK)<br><br>Owner: Gander International Airport Inc.<br>www.ganderairport.com<br>Province: Newfoundland and Labrador<br>Country: Canada | 9/11 Flight Diversions<br>August 2006 thwarted terrorism | Emergency response operations<br>Sheltering<br>Refugee services and processing<br>Continued "normal" airport operations for passengers, freight, and flight support Medical services<br>Communications<br>Logistics<br>  Staging relief<br>  Receiving relief<br>Provide physical facilities<br>Continuation of operations (COOP) | Terrorism<br>Catastrophe at major North American or European destination airport<br>Accidental failure of ATC system<br>Pandemic | *Beach head<br>*Command and control<br>*Emergency response operations<br>*Logistics<br>  *Staging relief<br>  *Receiving relief<br>*Sheltering<br>*Medical services<br>*Own-force protection<br>*Intelligence<br>*Communications<br>*Refugee services and processing<br>†Fiscal/administrative services<br>*Provide physical facilities<br>*Continuation of operations (COOP)<br>†Military operations<br>*Continued "normal" airport operations for passengers, freight, and flight support |

YQX and St. John's International Airport (YYT) are fascinating to examine in terms of transatlantic air traffic. Presently (2007 Feb 9, 1114 EST), 57 flights were closer to these two airports than to any other airport capable of landing and servicing large planes (Flightaware 2007). YQX stands proxy for all the reports receiving diverted and grounded flights, and YQX is the most extreme case as it was the airport attached to the smallest city of all the affected Canadian airports except Goose Bay in

Labrador. YQX's 9/11 experience was almost replicated in August 2006 when the "liquid bomb" threat to transatlantic airliners almost resulted in a diversion and grounding similar to 9/11 (Vey 2007; Armstrong 2006). This close call gave YQX, its nearby communities, and emergency response assets a chance to examine planning and preparedness progress since 2001 (Vey 2007).

Thirty-eight transatlantic airliners were diverted to and grounded at YQX, dumping 6,471 people into the airport at Gander, Newfoundland (Vey 2007). The Gander urban area had 11,234 inhabitants, a Salvation Army unit, and a Canadian Red Cross chapter in 2001 and now has one hotel, five motels, one bed and breakfast, and one campground. Ground transportation assets were limited. Fortunately from the point of view of **sheltering** and **logistics**, the citizens and hostelries opened their homes and rooms, and the Red Cross and Salvation Army functioned well until the last passengers left (Vey 2007; GIAA 2007). The effort was almost totally *ad hoc* but successful. There was no formal command, but a committee coordinated all activities (Vey 2007). All YQX facilities, government services, and tenant businesses cooperated in the response, prioritizing victim needs. Citizens of Gander helped for 9/11, and teachers, students, and services clubs were involved (Vey 2007). A community mobilization happened because of good relationships between YQX and its community. Routine airport operations were not compromised once the North American no-fly order was lifted on the third day.

If all transatlantic flights had been grounded in response to the thwarted "liquid bomb" terrorist threat at LHR in August 2006 (Armstrong 2006), North Atlantic air traffic control estimated about 60 planes would have gone to YQX (Vey 2007). The increase from 2001 probably reflects the recovery of passenger air travel since 2003 and the time of day.

Were YQX and the community better prepared in 2006 than in September 2001? "Yes. Most of the blankets and the like are still stored and we would be less surprised by the requirements of so many people arriving at one time." (Vey 2007) YQX has revised all emergency plans to optimize response to a similar incident in the future (Vey 2007).

### *Los Angeles International Airport (LAX) Faces Future Disasters*

LAX is included in this paper as a prime example of a proactive airport that is trying to build sound disaster management into its emergency plans. In 2005, LAX ranked 5th among world passenger airports and 8th for cargo (Infoplease.com 2007). Table 3 presents results for LAX.

LAX has advanced planning underway and nearly complete for joint, regional response to the likeliest types of disasters in Los Angeles. Issues of access, spaces, command and control, preparedness, and interoperability have been addressed, as well as pre-wiring and pre-designating spaces suitable for command facilities in physical plant planning (LAX, 2006a and b). LAX is scrambling to comply with revised FAA airport certification standards (FAA 2004). It can also be inferred that LAX is including in its planning the possibility of a biological incident.

**Table 3. Los Angeles International Airport.**

| Airport | Major Regional Threat(s) | Actual Response Uses and Mitigation Measures | Future Greatest Regional Threat(s) | Planned Responses and Mitigation Measures Based on Airport Plans and Documents |
|---|---|---|---|---|
| Los Angeles International Airport (LAX)<br><br>Owner: Los Angeles World Airports<br>http://www.la wa.org/lax<br>State: California<br>Country: USA | Earthquake<br>Watts Riots | Provide physical facilities<br>Command and control<br>Beach head<br>Emergency response operations<br>Logistics<br>  Staging relief<br>  Receiving relief<br>  Medical services<br>Own-force protection<br>Intelligence<br>Communications<br>Refugee services and processing<br>Fiscal/administrative services<br>Continuation of operations (COOP)<br>Continuation of business (COB)<br>Continuation of government (COG)<br>Military operations<br>Continued "normal" airport operations for passengers, freight, and flight support | Earthquake<br>Riots<br>Pandemic | *Beach head<br>*Command and control<br>*Emergency response operations<br>*Logistics<br>  *Staging relief<br>  *Receiving relief<br>*Sheltering<br>*Medical services<br>*Own-force protection<br>*Intelligence<br>*Communications<br>*Refugee services and processing<br>*Fiscal/administrative services<br>*Provide physical facilities<br>*Continuation of operations (COOP)<br>*Continuation of business (COB)<br>Continuation of government (COG)<br>†Military operations<br>*Continued "normal" airport operations for passengers, freight, and flight support |

Two overwhelming lessons jump out of the LAX case study. Advanced all-hazards assessment, training, and exercising are the keys to good preparedness, and plans are worthless unless backed up with financial and operational support. In concept, LAX's JCCs are intended to give the IC/UC a "warm start" prior to initiating a UC/IC or activating the EOC.

### London Heathrow International Airport (LHR) Prepares for Major Incidents

LHR is the world's preeminent international airport. It has vast visibility and economic importance to much of the world. In 2005, LHR ranked 3rd among world passenger airports and 17th for cargo (Infoplease.com 2007).

LHR has long been the focal point of aviation-related terrorist incidents and has cooperated with emergency management (LESLP 1999; LESLP 2004; London Resilience 2005; Hillingdon 2007). It is crucial for disaster relief efforts in Europe, Africa, and the Middle East, as well as potentially for North America and the Caribbean. It must prepare for the challenges of biological incidents. Table 4 summarizes LHR results.

LHR is more intimately connected to national and regional infrastructure than most airports. Its spaces, personnel, and communications capabilities make it suitable for a major role in any disaster. LHR and the surrounding agencies have planned, trained, and exercised for all hazards (LESLP 2004; London Resilience 2005; Hillingdon 2007). As a result of this activity, local government boundaries have been altered to ensure that LHR lies wholly within one London borough. This was done following lessons identified during emergency response events and exercises when LHR was dissected by the boundaries of 3 London Boroughs and 1 County Council (Rabjohn 2007).

| Table 4. London Heathrow International Airport. | | | | |
|---|---|---|---|---|
| **Airport** | **Major Regional Threat(s)** | **Actual Response Uses and Mitigation Measures** | **Future Greatest Regional Threat(s)** | **Planned Responses and Mitigation Measures Based on Airport Plans and Documents** |
| London Heathrow International Airport (LHR)<br><br>Owner: BAA<br>http://www.he athrowairport.c om/<br>County: London Borough of Killingdon<br>Country: UK | Pan Am 103 Terrorism<br>August 2006 thwarted terrorism | Staging relief to Europe, Africa, or SW Asia<br>Continued "normal" airport operations for passengers, freight, and flight support<br>Command and control<br>Provide physical facilities<br>Emergency response operations<br>Own-force protection<br>Intelligence<br>Communications<br>Logistics<br>Refugee services and processing<br>Fiscal/administrative services<br>Continuation of operations (COOP)<br>Continuation of business (COB)<br>Continuation of government (COG)<br>Military operations | Terrorism<br>Pandemic | †Staging relief to Europe, Africa, or SW Asia<br>†Beach head<br>*Command and control<br>*Emergency response operations<br>†Logistics<br>* Medical services<br>*Own-force protection<br>*Intelligence<br>*Communications<br>†Refugee services and processing<br>*Provide physical facilities<br>*Continuation of operations (COOP)<br>*Continuation of government (COG)<br>†Military operations<br>*Continued "normal" airport operations |

The British version of NIMS is a three tier structure of Strategic (Gold), Tactical (Silver) and Operational (Bronze) Command. Emergency Operations Command for LHR at the Gold level is coordinated by the Aviation Security Operational Command Unit (SO18), a command and control and special operations unit of the London Metropolitan Police (Answers.com 2007). In general British police terminology, it is a Special Operational Command Unit (SOCU) operating at LHR in the London Borough of Hillingdon. It provides policing and security for LHR and it supports the physical and organization infrastructures for unified commands foreseen by LHR's EOPs (LESLP 2004; Hillingdon 2007; Answers.com 2007).

Rabjohn (2007) summarizes the central, positive role of LHR in emergency management and disaster preparedness in the UK and indeed the world with the example of Victim and Recovery and Identification Team (VRIT). First created at LHR in 1987, they have been used repeatedly in major disasters worldwide. They are similar to Disaster Mortuary Teams (D-MORTs) in the U.S. Heathrow's VRIT deployed to New York after 9/11. Mutual aid pacts give a VRIT capability to deploy anywhere disaster involves British citizens.

LHR's experience since 1987 demonstrates that if an airport prepares thoroughly for one major threat using an all-hazard approach, total preparedness is enhanced for everything. If a major airport works out an all-hazard preparedness and response system that works, it is readily transportable to disasters far beyond the borders of that airport's country. A lesson is not really **learned** until it is fully implemented, trained, tested in exercises, practiced in reality, and subjected to the continuous improvement cycle.

## *Memphis International Airport (MEM), FedEx, and the National Guard*

MEM, containing FedEx Corporation's main office and primary world operating base, has become the world's number one cargo airport (Infoplease.com; MSCAA 2006; MSCAA 2007). MEM engages in intensive cooperative disaster preparedness with FedEx and the large co-located National Guard (NG) contingent. This was exemplified in the past three years by the construction of a new NG facility to free space adjacent to FedEx for that company's expansion (MSCAA 2006). MEM has much reserve capacity and few barriers to quick operational expansion in an emergency.

The MEM-FedEx-NG nexus creates extraordinary capabilities for dealing with a disaster in Memphis. MEM has the most extensive and capable ground support equipment for logistics in the world, and it probably has the greatest enclosed space in buildings of any airport except perhaps those at aircraft factories. Table 5 summarizes results for MEM/FedEx/NG.

**Table 5. Memphis International Airport.**

| Airport | Major Regional Threat(s) | Actual Response Uses and Mitigation Measures Described in Documents | Future Greatest Regional Threat(s) | Planned Responses and Mitigation Measures Based on Airport Plans and Documents |
|---|---|---|---|---|
| Memphis International Airport (MEM)<br><br>Owner: Memphis-Shelby County Airport Authority<br>www.mscaa.com<br>State: Tennessee<br>Country: USA | Mississippi River floods | Logistics<br>Military operations<br>Military support<br>Emergency response operations<br>Provide physical facilities<br>Continuation of operations (COOP)<br>Continuation of business (COB)<br>Continued "normal" airport operations | Earthquake<br>Mississippi River flood<br>Pandemic | †Staging relief to North America, Caribbean, Central America, or South America<br>*Military support<br>†Beach head<br>†Command and control<br>†Emergency response operations<br>*Logistics<br>*Own-force protection<br>†Intelligence<br>†Communications<br>†Provide physical facilities<br>†Continuation of operations (COOP)<br>†Continuation of business (COB)<br>*Military operations<br>*Continued "normal" airport operations |

Although MEM is not yet a major international gateway, it is a large domestic hub. As such, it faces a special role in dealing with a biological event. Flexible, strong response capabilities must be built into the MEM-FedEx- NG complex.

## Discussion

Airports are an essential component of emergency management. To maximize usefulness, it is necessary to act quickly to prepare them for key disaster management roles. The hidden need is financial support for capacity building. Airports can be useful assets in a variety of incidents, and the incident will determine how assets are deployed and employed as strategic goals develop. With proper development, resiliency can be created in the context of ready and capable facilities and procedures even though recovery efforts at an airport facility are complex.

### Opportunities for Action

➢ *FAA and federal resources.* By being the primary source of airport certification and funding, the Federal Aviation Administration (FAA) can drive initiatives towards improving emergency capabilities. Pre-wiring for communications and intelligence features appears not to be in the 2004 FAA mandate (14CFR, Part 139), though "pre-wiring" is feasible and affordable.

➢ *State influence or power.* States can be expected to justify new investments in capabilities at airports in terms of economic development, homeland security, and emergency management. It is conceivable that the National Guard may be the vehicle funding improvements at dual-use airports.

➢ *New construction, reconstruction, and retrofitting.* One possibility for funding physical preparedness and mitigation measures would be a set-aside program in airport construction grants and loans. This constant rewiring and upgrading of communications systems presents another opportunity for improvement.

➢ *Organizational and fiscal policy changes.* A change in organizational and fiscal policies can pay large dividends in preparedness and overall effectiveness of airports in disasters. Well-prepared airports have learned that success in supporting emergency responses pays long-term dividends in terms of public confidence. Even with all the problems during the Katrina response, MSY has clearly benefited from its role (Woodruff 2007).

➢ *Paying for the system.* There is consensus worldwide that air travelers and shippers should pay for the extra costs of airport-related improvements through surcharges or taxes, but a new mechanism is needed to fund the non-airport-related portion of capacity building.

➢ *JCC in the airport.* Preparing the airports better to host and support JCCs will improve the effectiveness of information management. Coordinated actions maximizing use of an airport's assets should be the way of the future.

➢ Sustaining advantages from actions – managing the improved system. The existing FAA Airport Certification Manual approach can promote and help sustain adherence to improved preparedness at airports, but something more may be needed. A standardized certification similar to Leadership in Energy and Environmental Design (LEED) certification. (USGBC 2007) could provide a benchmark.

### Summary and Conclusions

The scope of facilities and incidents used in this paper demonstrates the need for flexibility, scalability, and imagination in designing and executing pre-wired capabilities at airports. The case studies contained herein illustrate the ability of progressive airports to prepare for enhanced roles in disaster management. New Orleans and Katrina may seem to dominate this study; they are certainly the most thoroughly tested of a major airport during a regional disaster. However, the other cases offer important insights into the productive and successful use of airports for disaster response.

The primary thesis - that sound emergency management facilities, procedures, and organization should be built into airports and their operations to better serve citizens, optimize asset utilization, and facilitate airport resiliency - is supported by the findings of the case studies.

The thrust of most recommendations stemming from this study promote joint preparedness, interoperability, and coordination. The single most important recommendation is that increased public funding is needed to build better command and control capabilities for disasters. The costs will be small relative to the existing and projected future investments in homeland security, small compared to the economic impact of airports, and small compared to the potential savings in the response, recovery, remediation, and reconstruction phases of disasters. Building sound emergency management into airports should be among the most cost-effective all-hazard mitigation measures open to society today. Indeed, it may be the single most cost-effective form of mitigation.

## References

Answers.com. (2007) "Aviation security operational command unit," http://www.answers.com/topic/aviation-security-operational-command-unit (Jan. 15, 2007).

Armstrong, D. (2006) "Airlines take another hit: Bomb plot in Britain comes as U.S. carriers were starting to recover from years of losses." *San Francisco Chronicle,* Aug. 11, A1, http://www.sfgate.com (Jan. 27, 2007).

Chang, S. E., Ericson, D., and Pearce, L. (2003) *Airport closures in natural and human-induced disasters: Business vulnerability and planning,* Government of Canada, Office of Critical Infrastructure Protection and Emergency Preparedness, Ottawa.

Federal Aviation Administration (FAA). (2004) 14 CFR Part 139, Certification of Airports, Final Rule. (PowerPoint, slides 61-68), http://www.faa.gov/airports_airtraffic/airports/airport_safety/part139_cert/media/part 139_presentation.ppt#256 (Feb. 8, 2007).

Fenton, B. (2002) "Nation's airline traffic is grounded." *The Telegraph,* Dec 9 http://www.telegraph.co.uk/news/main.jhtml?xml=/news/2001/09/12/wfent12.xml (Jan. 26, 2007).

Flightaware.com. (2007) "Live flight tracker," http://flightaware.com/live/ (Feb. 9, 2007, 1114 EST).

Gander International Airport Authority (GIAA). (2007) "Gander International Airport," http://www.ganderairport.com (Jan. 29, 2007).

Hillingdon, London Borough of. (2007) "Civil Protection Service," http://www.hillingdon.gov.uk/environment/emergency_plan/what_we_do.php (Jan. 15, 2007).

Infoplease.com. (2007) "World's 30 busiest airports by passengers and cargo, 2005," http://www.infoplease.com/ipa/A0004547.html (Feb. 9, 2007).

London Emergency Services Liaison Panel (LESLP). (1999) *Major incident procedure manual 1999/2000*. LESLP, London, http://www.leslp.gov.uk (Jan. 15, 2007).

London Emergency Services Liaison Panel (LESLP). (2004) *Major incident procedure manual*. LESLP, London, http://www.leslp.gov.uk/LESLP_Man.pdf (Jan. 15, 2007).

London Resilience. (2005) *Strategic emergency plan: An overview of the Strategic London response to emergencies; summaries and highlights of pan-London arrangements* (ver. 2.1), London Prepared, London, http://www.londonprepared.gov.uk/londonsplans/emergencyplans/emergplan.pdf (Jan. 15, 2007).

Los Angeles International Airport (LAX). (2006a) *Draft ASAC all-risk / hazard airport security plan*, LAWA, Los Angeles.

Los Angeles International Airport (LAX). (2006b) *Draft concept of operations— ConOps2*, LAWA, Los Angeles.

Melancon, C., and Jefferson, W. J. (20060 "Additional views presented by the Select Committee on behalf of Rep. Charlie Melancon [and] Rep. William J. Jefferson," (2006) *a failure of initiative: the final report of the select bipartisan committee to investigate the preparation for and response to Hurricane Katrina*, Appendix B, U.S. House of Representatives, Washington, http://katrina.house.gov/ minority_views.doc (Feb. 3, 2007).

Memphis-Shelby County Airport Authority (MSCAA). (2006) *MSCAA Annual Report 2005*, MSCAA, Memphis, http://www.mscaa.com/AnnualReport2005.pdf (Jan. 21, 2007).

Memphis-Shelby County Airport Authority (MSCAA). (2007) "About Memphis Airport," www.mscaa.com (Feb. 7, 2007).

Miller, H., McNamara, J., and Jui, J. (2005) *Hurricane Katrina - after action report, OR-2 DMAT, New Orleans Airport August 31 to September 10, 2005*. OR-2 DMAT, Portland (OR), http://www.disastersrus.org/katrina/20051209101252-51802.pdf (Feb. 8, 2007 and as footnote 262 in House Katrina report).

National Commission on Terrorist Attacks upon the United States. (2004) *The 9/11 Commission report: Final report of the National Commission on Terrorist Attacks Upon the United States* (Official Government Edition), GAO, Washington, http://www.gpoaccess.gov/911/index.html (Feb. 3, 2007).

New Orleans Aviation Board (NOAB). (2007) "Louis Armstrong New Orleans International Airport, Statistics," http://www.flymsy.com (Jan. 27, 2007)

Nossiter, A. (2007) "Senators at Louisiana hearing criticize federal recovery aid." *N. Y. Times,* Jan. 30, A1.

Rabjohn, A. (2007) Personal communication, Feb. 15.

Rodriguez, M. (2006) "Lessons learned by Louis Armstrong New Orleans International Airport during Hurricane Katrina," undated PowerPoint presentation, 26 slides.

Rodriguez, M. (2007) Personal communication, Jan. 29.

Select Bipartisan Committee to Investigate the Preparation for and Response to Hurricane Katrina. (2006) *A failure of initiative: The final report of the Select Bipartisan Committee to Investigate the Preparation for and Response to Hurricane Katrina,* U.S. House of Representatives, Washington, http://katrina.house.gov/full_katrina_report.htm (Feb. 3, 2007).

U.S. Green Building Council. (2007) "Leadership in energy and environmental design (LEED)," http://www.usgbc.org/DisplayPage.aspx?CategoryID=19 (Feb. 9, 2007).

Vey, G. (2007) Personal communication, Jan 31.

White House. (2002) *Homeland Security Presidential Directive/HSPD-5, Management of domestic incidents,* Feb. 28, White House, Washington, http://www.whitehouse.gov/news/releases/2003/02/20030228-9.html (Apr. 22, 2006).

Woodruff, M. (2007) Personal communication, Jan 28.

# Development of a Intercity Mode Choice Models for New Aviation Technologies

Senanu Ashiabor[1], Antonio Trani[2], Hojong Baik[3], Nicolas Hinze[4]

[1]Via Department of Civil Engineering, Virginia Polytechnic Institute and State University, Blacksburg, VA 24061; PH (540) 257-3830; FAX (540) 231-7352; Email: sashiabo@vt.edu
[2]Via Department of Civil Engineering, Virginia Polytechnic Institute and State University, Blacksburg, VA 24061; PH (540) 231-2362; FAX (540) 231-7352; Email: vuela@vt.edu
[3]Via Department of Civil Engineering, Virginia Polytechnic Institute and State University, Blacksburg, VA 24061; PH (540) 231-4418; FAX (540) 231-7352; Email: hbaik@vt.edu
[4]Via Department of Civil Engineering, Virginia Polytechnic Institute and State University, Blacksburg, VA 24061; PH (540) 231-2362; FAX (540) 231-7352; Email: nhinze@vt.edu

## Abstract

A family of nested logit random utility models was developed to study intercity mode choice behavior in the United States. The models were calibrated using a nationwide revealed preference survey (1995 American Travel Survey) and two stated preference surveys conducted by Virginia Tech at selected airports in the U.S. The focus of this paper is on the ability of the models to estimate market share for the new category of Very Light Jet aircraft used in on-demand air taxi services. Analysis was performed to compare the stated preference surveys and the American Travel Survey within the same random utility framework. The main explanatory variables in the utility functions are travel time and travel cost stratified by household income. The model has been integrated into a large-scale computer software travel demand framework called the Transportation Systems Analysis Model to estimate nationwide intercity travel demand flow between 3,091 counties in the U.S., 443 commercial service airports and more than 3,000 general aviation airports in the U.S. A pared down version of the model will be integrated into the National Strategy Simulator that the FAA uses for strategic level planning the aviation system.

**Introduction**

Over the past few years Virginia Tech Air Transportation Systems Laboratory (VT ATSL) has been developing a multi-modal transportation decision support model to help policy makers' better plan and manage the nation's transportation system. The model is called the Transportation Systems Analysis Model (Baik and Trani 2005). NASA, FAA and the Joint Program Development Office (JPDO) are a few Federal agencies that have used demand forecasts from the model in planning activities. The core of the model is the classical four-step transportation planning technique made up of trip generation, trip distribution, mode choice and trip assignment. The mode choice portion of the model produces demand forecasts of the number of automobile, commercial air and Very Light Jet (VLJ) trips between 3091 counties in the U.S. In addition an airport choice model, integrated into the mode choice model, produces demand forecasts of number of commercial air trips between 443 commercial service airports in the U.S. and the number of VLJ trips between more than 3000 airports in the U.S. This paper focuses on the mode choice portion of the travel demand forecasting model.

The term Very Light Jets (VLJ) refers to jet-powered aircraft weighting less than 10,000 lbs. Examples of aircraft in this category are Eclipse Aviation 500, Cessna Mustang and Adams 700. TSAM was developed during NASA's Small Aircraft Transportation Systems program and at that time the vehicles were referred to as SATS aircraft. Since the end of the program the term VLJ has been used more in the aviation community. We use both terms: VLJ and SATS synonymously in this paper.

The goal of using TSAM as a national level planning tool drove certain aspects of the model structure and data requirements. It was important to have a multi-modal model so the impact of policies directed at one mode on other transportation modes could be studied. The need for a nationwide model meant national level databases were required for calibration. The 1995 American Travel Survey (ATS) was used because it was the largest nationwide survey of U.S. travelers available to the general public at the time the model was developed.

The ATS is a nationwide household travel survey of the U.S. residents conducted in 1995. The survey has more than 556,000 household trip records. Though the sample is large the survey has certain limitations. Privacy legislation in the U.S. limits the level of geographical information about trip origins and destinations available in the public version of the survey. The smallest geographical units in the survey are metropolitan statistical areas (MSA's). This limits the ability to develop credible estimates of travel times and other trip related variables from the survey. In addition, there is no information about origin and destination airports in the survey. The only airport related information provided is the drive time to the origin airport. Thus the ATS is a large survey of travelers' travel choices with very little information about the level of service variables that might have driven those choices.

The VT ATSL conducted four personal travel choice surveys to measure travelers' response to VLJ's. These surveys contain some amount of bias due to the small sample size (compared to the ATS), the limited geographical scope and the fact that they are not random household surveys of the U.S. population. However, during these surveys relevant level of service variables such as travel time, travel cost, and airport information was collected to enable more credible estimation of travel times, costs and modeling of airport choice behavior. This paper calibrates mode choice models using the Virginia Tech surveys and compares them with the ATS mode choice model in TSAM.

The next section is a review of the literature on past attempts to develop national level mode choice models and existing state-of-the-art in mode choice modeling. It is followed by a description of the methodology used to develop the mode choice models from the Virginia Tech surveys. We then present the outputs of the calibrated models and an analysis of the results. The final section is made up of comments and recommendations for further work to improve the choice model.

## Literature Review

### *National Level Travel Demand Models*

The use of early logit-type mode choice models in transportation planning began in the early 70's. Table 1 is a summary of five national-level intercity mode choice models of the U.S. found in the literature to date. The models were developed by Stopher and Prashker (1976), Grayson (1982), Morrison and Winston (1985), Koppelman (1990), and Ashiabor et. al. (2007). All these models were calibrated with some version of the Census Bureau National Travel Surveys. The first two were multinomial logit, the third and fourth were nested logit models. In the fifth model by the authors, both nested and mixed logit models were developed. The major variables used in all the models were travel time and cost. Separate models were calibrated for business and non-business trip purpose in all cases. Overall the modeling attempts demonstrate that disaggregate modeling procedures can be effectively applied to estimate mode choice models with certain limitations.

The absence of level of service variables in the National Travel Surveys forced modelers to create synthetic level of service variables from external sources such as the Official Airline Guide (OAG 2000) and the BTS Airline Origin and Destination Survey usually referred to as DB1B (BTS, 2000). Several assumptions have to be made about location of trip ends in the ATS surveys, that affect the credibility of access and egress cost and time estimates. The high level of aggregation introduced in the creating synthetic travel times makes model results subject to aggregation bias. We believe this is a key reason why all the models prior to Ashiabor et al (2007) restrict their geographical scope of analysis to only MSA's. The restriction helps provide more credible level of service estimates but renders the models almost useless for application to non-MSA areas. It is even more difficult to identify both the airport used by the traveler and the alternative airports that were considered in the National Travel Surveys. Due to this issue none of the models prior to Ashiabor et al (2007) directly incorporates airport choice. The absence of airport choice is a crucial deficiency since demand through airports is a key component in analyzing the National Airspace System. Koppelman and Hirsh (1986) highlight the data requirements that would be needed for researchers and practitioners to develop accurate and useful intercity travel demand models. However, to date there does not seem to be any attempt by either the Census Bureau, or the Bureau of Transportation statistics to collect such data.

Lewe et al (2003) developed a nation level model using an agent-based approach to model traveler's response to Very Light Jets (during NASA's SATS program). Travelers in Lewe's model are agents that seek to complete trips comfortably and safely with less travel time. The mode choice portion of Lewe's model uses a multinomial logit model. Given the documented weaknesses and limitations of the multinomial logit models for forecasting, such a model might be severely constrained for modeling travelers responses (Koppelman, 1990).

The mode choice models, presented in this paper, extend the work of the models mentioned above along three dimensions. The TSAM mode choice model used both MSA and non-MSA information in the ATS so the model can be applied at the national level. Secondly, a heuristic approach is introduced to implement airport choice modeling to enable the model estimate market share and by extension demand through airports, making the applied model more useful to policy makers. Thirdly, synthetic travel times are aggregated from the county level for the whole U.S. in order to justify nationwide applicability of the model. TSAM is one of the few large scale multi-mode intercity choice models to combine mode choice and airport choice in the U.S. at the county level in a single scheme.

### Review of Logit Models

McFadden (1973) conditional logit model (later known as the multinomial logit) was one of the earliest logit models applied in the transportation literature. He derived it based on Luce's (1959) utility maximizing axiom which states that "*If one is comparing the two alternatives according to some algebraic criterion, say preference, this comparison should be unaffected by the addition of new alternatives or the subtraction of old ones*" – *Luce 1959*. The multinomial logit model assumes a Gumbel distribution and a random sample that is Independent and Identically Distributed (IID). This implies that alternatives being considered are independent of each other and have the same variance. Imposing the IID assumption yields a simple model with a probability of the form $P(i) = e^{V_i} / \sum_{j=1}^{J} e^{V_j}$. It is obvious that for any two alternatives $s$ and $t$ the ratio of their probabilities $P(s)/P(t) = e^{V_s} / e^{V_t}$ is independent of the presence of any other alternatives in the model as postulated in Luce's axiom. Despite early successes in using the multinomial logit model for empirical studies it came under heavy criticism due to the IIA property and its restrictive substitution patterns (Ben-Akiva and Lerman 1985). In order to provide more flexible empirical models either the Independence or Identical Distribution assumptions needs to be relaxed while maintaining the closed form model.

The nested logit model relaxes the independence assumption by grouping alternatives that are similar into nests (McFadden 1978; Daly and Zachary 1978). A plethora of models have been developed since the nested logit model was developed but McFadden derived a General Extreme Value (GEV) model that is an overarching framework from which all the models that relax either assumption can be derived. Mixed logit models go a step further to relax both the Independence and Identically Distribution assumptions. The earliest mixed models were by Byod and Mellman (1980) and Cardell and Dunbar (1980). The mixed models developed since include Brownstone and Train (1999), Bhat and Castelar (2002) and Hess and Polak (2005).

Table 2 summarizes the broad classes of logit models used in transportation planning in terms of the assumptions they attempt to relax. A detailed review of logit models in the transportation literature is available in Ashiabor (2007).

We use nested logit models for this paper because initial mixed logit models calibrated for TSAM did not show substantial gains in predictive power over the nested logit model. The variables used are travel time and travel cost segregated by household income.

## Model Development

### *Form of Logit Models*

The mode choice models presented are developed based in random utility theory (Train 2003). The random utility framework considers individuals as utility maximizing entities. When faced with a choice set of $J$ alternatives, each individual $n$ is assumed to attach a utility $U_{nj}$ to the $j^{th}$ alternative that is a composite of the individuals attributes and attributes of the alternative. The modeler has information about certain attributes of the individual $a_n$ and attributes of each alternative $x_{nj}$ " $j$ that are assumed to factor in the individual's decision making process. The utility of the individual can be framed as $U_{nj} = V_{nj} + \epsilon_{nj}$ where $\epsilon_{nj}$ is an error term representing additional information $U_{nj}$ known to the individual but not the modeler. The modeler observes $V_{nj}$ usually referred to as the representative utility and treats the term $\epsilon_{nj}$ " $j$ as random hence the name random utility. The probability that an individual selects a specific alternative $i$ from the $J$ alternatives is written as $P_{ni} = \text{Pr} ob\, U_{ni})U_{nj}$ " $j$ „ $i$). Implying individuals select the alternatives from the choice set that give them maximum utility hence, the term utility maximization. The probability can be computed as $P_{ni} = Prob\, V_{ni} + \epsilon_{ni})V_{nj} + \epsilon_{nj}\, K$ " $j$ „ $i$). The modeler seeks to specify an appropriate distribution for $V_{nj}$ that captures the individual's decision making behavior.

To derive the nested logit model assume the $J$ alternatives are further portioned into $K$ nests. We can decompose the utility as $U_{nm} = W_{nm} + Y_{nj} + \epsilon_{nj}$ where $W_{nm}$ represents the utility associated with variables in the $m^{th}$ nest and $Y_{nj}$ is the utility linked to the $j^{th}$ alternative in the nest. It can be shown that probability $P_{ni}$ of an individual $n$ selecting alternative $i$ has the form in Equation 1.

$$P_{ni} = \left[ \frac{e^{Y_{ni}/\lambda_m}}{\sum_{j.M} e^{Y_{nj}/\lambda_m}} \right] \left[ \frac{e^{W_{nm}+\lambda_m I_{nm}}}{\sum_{l=1}^{M} e^{W_{nl}+\lambda_l I_{ni}}} \right] \tag{1}$$

where the terms in the brackets are the marginal and conditional distributions respectively. $I_{nm} = \sum_{j.M} e^{Y_{nj}/\lambda_m}$ is referred to as the inclusive value or utility and $\lambda_m$ is called the inclusive coefficient. $l$ represents each of the $j$ alternatives in nest $M$. A more detailed version of the derivation is available in Train (2003).

As mentioned earlier the Virginia Tech Air Transportation Systems Laboratory (VT ATSL) has calibrated a nested logit model in the Transportation Systems Analysis Model (TSAM). Variables used in the model were travel time and travel cost. Both variables were categorized by household income of the traveler. There are five household income categories in the model: households earning less than $30,000, from $35,000 to $60,000, from $60,000 to $100,000, from $100,000 to $150,000 and households earning more than $150,000. The income groups are indexed 1 to 5 in ascending order in the model. The utility function in the nested logit model has the form in Equation 2.

$$U_{ij}^{kl} = a_j TT_{ij}^{kl} + b_j TC_{ij}^{kl} \tag{2}$$

Where $U_{ij}^{kl}$ is the utility of a traveler belonging to income household group $l$, making a trip between origin $i$, and destination $j$, using mode of transportation $k$. $TT_{ij}^{kl}$ is the door-to-door travel time of a traveler belonging to income household group $l$, making a trip between origin $i$, and destination $j$, using mode of transportation $k$. $TC_{ij}^{kl}$ is the door-to-door travel cost of a belonging to income household group $l$, making a trip between origin $i$, and destination $j$, using mode of transportation $k$. $a_j$'s are the coefficients related to travel time for each income group and $b_j$'s are the coefficients related to cost for each income group.

The form of the utility equation above was chosen to enable to model capture the value of time of travelers in the different income groups. Separate models were calibrated for business and non-business travelers. This model was calibrated with the 1995 ATS survey and a detailed discussion on the computation, creation of synthetic travel times and costs, and development of the airport choice model for the mode choice model in TSAM is available in Ashiabor (2007) and Ashiabor et al. (2007). Travel cost changes by income group because higher income travelers are more likely to either purchase unrestricted/first class tickets or buy their tickets at the last minute.

A key aim of TSAM was to forecast demand for Very Light Jet (VLJ) aircraft operating as an air taxi system in the U.S. In order to estimate demand for VLJ's using the above model we need both travel times and costs and an inclusive coefficient for VLJ's. The travel times and costs can be derived synthetically however the inclusive coefficient needs to be calibrated from a survey. Since the ATS has no information on VLJ's the inclusive coefficient in the model was selected by making iterative runs for the VLJ inclusive coefficient keeping the value between one and the commercial air coefficient value. A key motivation of conducting the VT ATSL conducted stated preference surveys was to collect information on traveler's response to VLJ in order to more credibly estimate the inclusive value coefficient. The forms of utility models in this analysis are different from that in Equation 2 due to reasons explain later in the paper.

### Development of VT Personal Travel Surveys

A *stated preference* survey is one in which respondents are asked to respond to a hypothetical question or scenario in contrast to a *revealed preference survey* in which respondents are asked about actions they have taken in the past. The VT ATSL conducted four revealed and stated preference surveys to elicit travelers' response to VLJ's operating as air taxis. During the *revealed preference* portion of the survey travelers were asked to provide information about the last trip they made in the past six months. They were asked about the main mode of transportation used for the trip, and access and egress modes if the main mode was by air, train or bus. Data on travel times and cost for all segments of the trip was collected with information about trip origins and destinations. Information about the locations and number of people survey is provided in Table 3. Based on these the door-to-door travel time and cost of the trip can be calculated. Additional information collected included travelers age, education, and household income information.

In the *stated preference* portion of the survey we used the trip origin and destination provided by the respondent to select suitable candidate general aviation (or VLJ) airports close to the trip ends. We then generate door-to-door travel times and costs needed to make the same trip using VLJ aircraft from the candidate airports. The respondent was then presented with the travel times and costs for both their last trip and the hypothetical VLJ trip and asked to indicate which mode of transportation they would use if faced with this new scenario in the future. Figure 1 is a screen shot of the survey page that had the hypothetical question for a trip from Blacksburg, VA to New York, NY. For the purpose of our survey we define suitable general aviation airports as public use airports with a paved runway that is more than 3000 feet in length.

VT ATSL conducted four such surveys at the 2004 Oshkosh Air Venture Fair at Wisconsin. The fair is an annual gathering for private aircraft owners, and it was selected to gain their reaction to SATS program. An online internet survey, and two surveys at Roanoke and Reagan National airports. Summary information about the surveys is provided in Table 3. All the surveys were designed by the VT ATSL. For the online survey, a consulting firm was paid to provide online participants with a link to an ATSL computer server on which they filled out the survey form. The other surveys were all administered by personnel of the VT ATSL team using laptops pre-loaded with a programmed version of the survey.

Automobile travel times in the survey were computed on the fly using Microsoft MapPoint software (Microsoft Corporation, 2004). MapPoint was also used to compute drive times to and from airports. VLJ flight times were derived using a flight profile program developed at VT ATSL, coded in Mathworks MATLAB software. The door-to-door travel time for a VLJ trip is the sum of the automobile travel times to the airports, the flight time, the processing times at the airport and a schedule delay of one hour. We assumed processing times of 30 and 25 minutes at the origin and destination VLJ airports based on current times at general aviation airports. Schedule delay is a measure of additional travel time travelers are forced to experience because flights are not scheduled at the time the traveler actually wants to depart (Tedodorovic 1998).

Automobile travel costs were computed as the product of the route distance and an assumed per-mile cost. We assumed 37 cents per mile if traveler drove and $1 per-mile plus $3 dollars if taxi was used. These values were judgmentally selected based on reviews of government per-mile rates and information from the American Automobile Association website. For VLJ flights the unit cost was fixed at $1.75 per-seat-mile based on a lifecycle cost model developed by the VT ATSL group. The flight cost is the product of the round trip great circle distance between the airports and the cost-per-seat-mile. The door-to-door travel cost for VLJ's is the sum of the flight costs and the automobile travel costs. The survey interface was coded in Microsoft Visual Basic.

### Model Calibration

Models presented were calibrated using datasets from the Roanoke and Reagan National airport. The first two surveys are not used due to the small sample size, and changes made to the survey questionnaire after the first two surveys. The Roanoke dataset had 266 and 269 business and non-business records respectively. The Reagan National airport survey had 327 and 208 business and non-business records respectively.

The form of the utility model used in calibrating the VT Surveys is shown in Equation 3. The subscripts have the same meaning as in Equation 2.

$$U_{ij}^k = a_j TT_{ij}^{kl} + bTC_{ij}^k \tag{3}$$

This is restricted form is slightly different from Equation 2 because initial attempts to calibrate the VT survey models with the utility structure in Equation 2 produced either counterintuitive signs or unacceptable p-values. We tested various schemes starting with a base model that had no differentiation by income group, with the utility form $U_{ij}^k = aTT_{ij}^k + bTC_{ij}^k$, and varied travel time over income, and then varied cost over income. The model with the best estimates is the form in Equation 3. This model has five $a_j$'s, for time (one for each income group) but only one $b$ for cost. In order to compare the models from the VT survey datasets with the TSAM/ATS model we recalibrated the nested logit model in TSAM using the utility equation shown in Equation 3. The current nested logit model in TSAM have the form in Equation 2, and all the signs of the coefficients of travel time and cost in that model are negative. That model has a good fit and has been applied to estimate demand for automobile, commercial air and SATS (Baik and Trani, 2005). The form of the ATS model in this paper was adopted to compare the ATS and Virginia Tech survey datasets.

The business models are presented in Table 4, and the non-business in Table 5. For each Table there is a revealed preference model for the ATS, and a revealed and stated preference model for each airport survey. We also calibrate models from the combined data of the two airports. The stated preference models are calibrated using the responses supplied during the survey. The models were calibrated using the MDC procedure SAS statistical software package (SAS Institute). The structure of the Nested Logit model is shown in Figure 2.

### Discussion of Results

The models are assessed along a number of dimensions. We check the sign and magnitude of coefficients, goodness of fit statistics (r-square, log-likelihood), and also compare the models to each other. First the Virginia Tech Survey models will be discussed and then compared with the ATS model. For the Virginia Tech Survey models all the calibrated travel time and cost coefficients have negative signs. Indicating an increase in either travel time or cost is perceived negatively by travelers. This is intuitive as individuals normally prefer faster and cheaper modes.

For the revealed preference business models (Table 4) the goodness of fit statistics show that the Roanoke airport (ROA) model is better than the Reagan National (DCA) airport model and the combined model. Note that the closer the log-likelihood is to zero the better the model. The case of the business stated preference the DCA model has the better r-square but the ROA model has a slightly better log-likelihood. The magnitude of the coefficients of the revealed preference surveys is consistent across the three models, with the ROA and the combined survey being closer in magnitude. The stated preference survey combined survey yields values that are completely different from the ROA and DCA surveys. The p-values indicate all the coefficients are significant however; the inclusive coefficients for all the models are very close to zero. Inclusive coefficients close to zero are usually an indication that a multinomial logit model would provide as good a fit as the nested logit formulation.

For the ATS business model the travel time coefficient for the first income group is positive. This is unintuitive because it implies individuals with income less than $30,000 prefer transportation modes that take more time if cost is held constant. All the other coefficients are negative, and the absolute magnitudes of the travel time coefficients increase with income. Given only one cost coefficient the trend implies that value of time increases with income. The ratio $a_j / b$ is a measure of the magnitude of an individual's value of time.

For the non-business revealed preference models (Table 5) the DCA model performs better than the ROA model. However, in terms of the magnitude of the coefficients those of the combined survey are much closer to the ROA survey. Once again the inclusive coefficients are all close to zero, indicating the nested logit is a poor choice to fit the data. The non-business stated preference model coefficients and fit statistics are very poor for the combined model. We however note that all the coefficients for the ATS non-business model are negative. The absolute values of the travel time also increase monotonically indicating increasing value of time with income as expected.

**Comments and Conclusions**

A key motivation for the analysis was to estimate more credible inclusive coefficient values for TSAM. The low inclusive coefficients for all the Virginia Tech Survey models seem to indicate the nested logit is not a good fit. However, the high values of inclusive coefficients for the ATS model indicate otherwise. We believe the conflicting result from the Virginia Tech Surveys is due to the small sample size. The fact that initial attempts to calibrate a model segregated over income for both travel time and cost failed for the survey further strengthens our conviction that the small sample size affects the ability of the survey data to recover the true model parameters. The large difference between the magnitude of the coefficients (for both travel time and cost) for the ATS survey and the Virginia Tech Surveys cast some level of credibility over the results.

Given the size of the sample size of the ATS and Virginia Tech Surveys, and also the fact that the ATS is a random survey conducted over the population of the

U.S. the coefficients from that model are more credible. The sample size was limited by available funds and personnel to conduct the surveys. Based on this we recommend using the ATS model and a much larger data collecting effort in the future to recalibrate the model and improve the Virginia Tech Survey models.

**Table 1: Major National Level Intercity Mode Choice Models for United States.**

| | Model Type | Data and Scope | Modes of Transportation | Variables in Utility Function | Market Segmentation |
|---|---|---|---|---|---|
| Stopher and Prashker (1976) | Multinomial Logit | Database: 1972 NTS<br>Scope: Trips that start and end in Metropolitan Statistical Areas (MSA)<br>**2,085** records from database | Automobile, Commercial Air, Bus, Rail. | Relative time, relative distance, relative cost, relative access-egress distance, departure frequency | Trip purpose (Business / Non-Business) |
| Alan Grayson (1982) | Multinomial Logit | Database: 1977 NTS<br>Scope: Trips that start and end in MSA's<br>Selected observations from database | Automobile, Commercial Air, Bus, Rail. | Travel time, travel cost, access time, and departure frequency | Trip purpose (Business / Non-Business) |
| Morrison and Winston (1985) | Nested Logit | Database: 1977 NTS<br>Scope: Trips that start and end in MSA's<br>**4,218** records from database | Automobile, Commercial Air, Bus, Rail. | Travel time, cost, party size, average time between departures | Trip purpose (Business / Non-Business) |
| Koppeleman (1990) | Nested Logit | Database: 1977 NTS<br>Scope: Trips that start and end in MSA's<br>Selected observations from database | Automobile, Commercial Air, Bus, Rail. | Travel time, cost, departure frequency, distance between city pairs, household income. | Trip purpose (Business / Non-Business) |
| Ashiabor et. al. (2007) | Nested Logit and Mixed Logit Models | Database: 1995 American Travel Survey<br>Scope: All trips irrespective of origin or destination type.<br>**402,295** records from database | Automobile, Commercial Air, Train, Very Light Jets. | Travel time, Travel Cost, Household Income, Region Type. | Trip purpose (Business / Non-Business)<br><br>Household Income |

**Table 2.  Classification of Forms of Logit Models (Ashiabor, 2007).**

|  | Restrictive IID Model | GEV Models | | Relax Dependence & ID Assumptions |
|---|---|---|---|---|
|  |  | **Relax ID assumption** | **Relax Dependence Assumption** |  |
| **Assumed Distribution** | Type I Extreme Value or Gumbel | Type I Extreme Value or Gumbel | Type I Extreme Value or Gumbel | Type I Extreme Value + Specified Distribution |
| **Independence Assumption** | Independent | Independent | Correlation between alternatives | Correlation between alternatives |
| **Identical Distribution Assumption** | Identically Distributed | Variance of unobserved factors differ over alternatives | Identically Distributed | Variance of unobserved factors differ over alternatives |
| **Model Type** | *McFadden's Conditional logit* | *Heteroskedastic Extreme Value (HEV)* | *Nested logit* | *Mixed logit* |

**Table 3. Summary of Virginia Tech Revealed/Stated Preference Survey Information**

| Location | Survey Instrument | Date | Sample Size |
|---|---|---|---|
| **2004 Oshkosh Air-Venture Fair, WI** | **Laptop:** Survey interface was designed and pre-installed on laptop | July 27, 2004 to August 8, 2004 | 116 |
| **Internet-based Survey** | **Web-based survey:** respondents are contacted by email and given link to website with survey | November 12, 2004 to November 14, 2004 | 115 |
| **Roanoke Airport, Roanoke, VA** | **Laptop:** Survey interface was designed and pre-installed on laptop | March 16, 2005 to March 24, 2005 | 535 |
| **Reagan National Airport, Arlington, VA** | **Laptop:** Survey interface was designed and pre-installed on laptop | May 16, 2005 to May 19, 2005 | 535 |

**Table 4: Business Nested Logit Models for ATS and Virginia Tech Surveys**

| SURVEY NAME: | ATS | | Virginia Tech Survey | | | | | |
|---|---|---|---|---|---|---|---|---|
| LOCATION: | Nationwide | | Roanoke Airport | | Reagan National Airport | | Roanoke and Reagan Airports | |
| **Revealed Prefrence Model** | Coefficient | P-value | Coefficient | P-value | Coefficient | P-value | Coefficient | P-value |
| Travel Time (< $30,000) | 0.0159 | <.0001 | -0.3397 | 0.005 | -0.2737 | 0.0061 | -0.2066 | <.0001 |
| Travel Time ($30,000 to 60,000) | -0.0264 | <.0001 | -0.2521 | 0.0002 | -0.3559 | <.0001 | -0.2653 | <.0001 |
| Travel Time ($60,000 to 100,000) | -0.0652 | <.0001 | -0.5612 | 0.0001 | -0.3550 | <.0001 | -0.3809 | <.0001 |
| Travel Time ($100,000 to 150,000) | -0.1388 | <.0001 | -0.3017 | <.0001 | -0.2802 | <.0001 | -0.2452 | <.0001 |
| Travel Time (> $150,000) | -0.1550 | <.0001 | -1.0630 | 0.0027 | -0.4787 | <.0001 | -0.5311 | <.0001 |
| Travel Cost | -0.0188 | <.0001 | -0.0044 | 0.0076 | -0.0018 | 0.0339 | -0.0022 | 0.0005 |
| Inclusive Coefficient | 0.3538 | <.0001 | 0.0633 | <.0001 | 0.0713 | <.0001 | 0.0918 | <.0001 |
| **Statistics** | | | | | | | | |
| R-Square (Adjusted Estrella) | 0.8294 | | 0.9372 | | 0.8762 | | 0.8691 | |
| Log-Likelihood | -71,885 | | -130 | | -207 | | -389 | |
| **Stated Preference Model** | | | Coefficient | P-value | Coefficient | P-value | Coefficient | P-value |
| Travel Time (< $30,000) | | | -0.2566 | 0.0246 | -0.2325 | 0.0137 | -0.0358 | 0.0234 |
| Travel Time ($30,000 to 60,000) | | | -0.4719 | 0.0025 | -0.3608 | 0.0003 | -0.2390 | <.0001 |
| Travel Time ($60,000 to 100,000) | | | -0.5400 | 0.0004 | -0.4285 | <.0001 | -0.1386 | 0.0001 |
| Travel Time ($100,000 to 150,000) | | | -0.3916 | 0.0037 | -0.3570 | <.0001 | -0.1935 | 0.0004 |
| Travel Time (> $150,000) | | | -0.6663 | 0.0062 | -0.5617 | <.0001 | -0.3143 | <.0001 |
| Travel Cost | | | -0.0098 | 0.0002 | -0.0062 | <.0001 | -0.0072 | <.0001 |
| Inclusive Coefficient (Auto) | | | 0.0674 | 0.0009 | 0.0840 | <.0001 | 0.0402 | <.0001 |
| **Statistics** | | | | | | | | |
| R-Square (Adjusted Estrella) | | | 0.8747 | | 0.8868 | | 0.7164 | |
| Log-Likelihood | | | -219 | | -262 | | -428 | |

**Table 5: Non-Business Nested Logit Models for ATS and Virginia Tech Surveys**

| SURVEY NAME: | ATS | | Virginia Tech Survey | | | | | |
| --- | --- | --- | --- | --- | --- | --- | --- | --- |
| LOCATION: | Nationwide | | Roanoke Airport | | Reagan National Airport | | Roanoke and Reagan Airports | |
| | Coefficient | P-value | Coefficient | P-value | Coefficient | P-value | Coefficient | P-value |
| **Revealed Preference Model** | | | | | | | | |
| Travel Time (< $30,000) | -0.1216 | <.0001 | -0.2555 | 0.002 | -0.0659 | 0.4887 | -0.1997 | 0.0007 |
| Travel Time ($30,000 to 60,000) | -0.1340 | <.0001 | -0.4096 | 0.001 | -0.6527 | 0.0013 | -0.4767 | <.0001 |
| Travel Time ($60,000 to 100,000) | -0.1676 | <.0001 | -0.2762 | 0.0011 | -0.6469 | 0.0008 | -0.3706 | <.0001 |
| Travel Time ($100,000 to 150,000) | -0.2159 | <.0001 | -0.4395 | 0.0087 | -0.9323 | 0.0253 | -0.6667 | 0.0003 |
| Travel Time (> $150,000) | -0.3185 | <.0001 | -0.4229 | 0.1092 | -0.4277 | 0.0011 | -0.4561 | 0.0002 |
| Travel Cost | -0.0262 | <.0001 | -0.0037 | 0.0097 | -0.0087 | 0.0039 | -0.0044 | <.0001 |
| Inclusive Coefficient | 0.4255 | <.0001 | 0.0292 | 0.0004 | 0.0213 | 0.0012 | 0.0315 | <.0001 |
| **Statistics** | | | | | | | | |
| R-Square (Adjusted Estrella) | 0.9754 | | 0.8221 | | 0.8707 | | 0.8378 | |
| Log-Likelihood | -36,128 | | -136 | | -132 | | -281 | |
| **Stated Preference Model** | | | | | | | | |
| Travel Time (< $30,000) | | | -0.2175 | 0.0092 | -0.0050 | 0.9538 | -0.0955 | 0.008 |
| Travel Time ($30,000 to 60,000) | | | -0.2328 | 0.0016 | -0.8415 | 0.0001 | -0.0862 | 0.0001 |
| Travel Time ($60,000 to 100,000) | | | -0.1764 | 0.0011 | -0.6595 | 0.0007 | -0.2918 | <.0001 |
| Travel Time ($100,000 to 150,000) | | | -0.2459 | 0.0031 | -0.5120 | 0.0163 | -0.2679 | <.0001 |
| Travel Time (> $150,000) | | | -0.1370 | 0.1377 | -0.3755 | 0.0007 | -0.1529 | <.0001 |
| Travel Cost | | | -0.0076 | <.0001 | -0.0127 | <.0001 | -0.0056 | <.0001 |
| Inclusive Coefficient | | | 0.0586 | <.0001 | 0.0280 | 0.0002 | 0.0264 | <.0001 |
| **Statistics** | | | | | | | | |
| R-Square (Adjusted Estrella) | | | 0.7355 | | 0.7922 | | 0.6451 | |
| Log-Likelihood | | | -197 | | -199 | | -686 | |

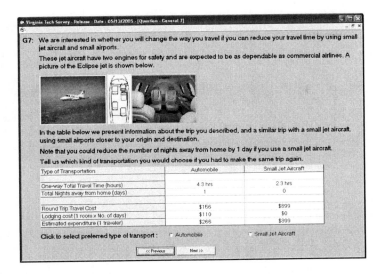

Figure 1. A page from VT Stated and Revealed Preference Survey for a Trip from Blacksburg, VA to New York, NY using Very Light Jet Aircraft.

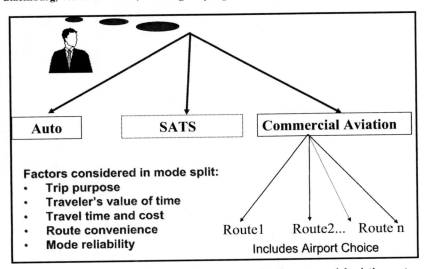

Figure 2. Structure of Nested Logit Model (Each route under the commercial aviation nest represents a trip between an Origin-Destination pair through one Candidate Airport at the Origin and another at the Destination).

## REFERENCES

Ashiabor, S. (2007). *Modeling Intercity Mode Choice and Airport Choice in the U.S.*, Ph.D. Dissertation, Virginia Tech.

Ashiabor, S., Baik, H., and Trani, A. (2007). Logit Models to Forecast Nationwide Intercity Travel Demand in the U.S., Accepted for publication In *Transportation Research Record: Journal of the Transportation Research Board*.

Ben-Akiva, M., and Lerman, S. (1985). Discrete Choice Analysis: Theory and Application to Travel Demand, MIT Press, Cambridge, MA.

Bhat, C., and Castelar, S. (2002). A unified mixed logit framework for modeling revealed and stated preferences: Formulation and application to congestion pricing analysis in the San Francisco Bay area, *Transportation Research Part B*, Vol. 36, pp 577-669.

Brownstone, D., and Train, K. (1999). Forecasting new product penetration with flexible substitution patterns, *Journal of Econometrics*, Vol. 89, pp 109-129.

Bureau of Transportation Statistics (2000). *Airline Origin and Destination Survey (DB1B)*, http://www.transtats.bts.gov/.

Bureau of Transportation Statistics (1995). *American Travel Survey: An overview of the survey design and methodology*, Bureau of Transportation Statistics.

Byod, J.H., and Mellman, R.E. (1980). The effect of fuel economy standards on the U.S. automotive market: An hedonic demand analysis, *Transportation Research Part A*, Vol. 14A, pp 367-378.

Cardell, N.S., and Dunbar, F.C. (1980). Measuring the societal impacts of automobile downsizing, *Transportation Research Part A*, Vol. 14A, pp 423-434.

Daly, A., and Zachary, S. (1978). Improved multiple choice models, In *Determinants of Travel Choice* (Edited by Hensher, D., Dalvi, M.), Saxon House, Sussex.

Grayson, A. (1982). Disaggregate model of mode choice in intercity travel, In *Transportation Research Record: Journal of the Transportation Research Board, No. 385*, pp 36-42.

Baik, H., and Trani, A. (2005). *A Transportation Systems Analysis Model (TSAM) to study the impact of the Small Aircraft Transportation System (SATS)*, http://www.atsl.cee.vt.edu/sds_papers.htm#TSAM.

Hess, S., and Polak, J.W. (2005). Mixed logit modeling of airport choice in multi-airport regions, *Journal of Air Transport Management*, Vol. 11, pp 59-68.

Koppelman, F.S. (1990). Multidimensional Model System for Intercity Travel Choice Behavior, In *Transportation Research Record: Journal of the Transportation Research Board, No1241*, pp 1-8.

Koppelman, F.S. and Hirsh, M. (1986). Intercity passenger decision making: conceptual structure and data implications, In *Transportation Research Record: Journal of the Transportation Research Board, No. 1085*, pp 70-75.

Lewe, J., Upton, E., Marvis, D., and Schrage, D. (2003). An Agent-Based Framework for Evaluating Future Transportation Architectures, AIAA 3rd Annual Aviation Technology, Integration and Operations (ATIO) Forum, Nov. 17-19.

Luce, D. R. (1959). *Individual Choice Behavior*, New York: Wiley.

McFadden, D. (1973). Conditional logit Analysis of Qualitative Choice Behavior, in P. Zarembka, *Frontiers in Econometrics*, Academic Press: New York, pp 105-142.

McFadden, D. (1978). Modeling the choice of spatial location, In *Spatial Interaction Theory and Planning Models* (Edited by Karlqvist, A., Lundqvist, L., Snickars, F., and Weibull, J.), North-Holland, Amsterdam, pp 75-96.

Microsoft Corporation (2004). *MapPoint: Business Mapping and Data Visualization Software*, CD-ROM.

Morrison, S. and Winston, C. (1985). An Econometric Analysis of the Demand for Intercity Passenger Transportation, *Research in Transportation Economics*.

Official Airline Guide (OAG), CD-ROM, 2000.

SAS Institute, http://www.sas.com, version 9.1.3.

Stopher, P., and Prashker, J. (1976). Intercity Passenger Forecasting: The Use of Current Travel Forecasting Procedures, *Transportation Research Forum: Annual Meeting Proceedings*, pp 67-75.

Teodorovic, D. (1998). *Airline Operations Research*, Gordon and Breach Science Publishers, New York.

Train, K.E. (2003). *Discrete Choice Methods with Simulations*, Cambridge University Press.

# Measuring Pre-Flight Travel Time in the Air Journey

## A.K. Goswami[1], J.S. Miller[2], and L.A. Hoel[3]

[1]Graduate Research Assistant; University of Virginia & Virginia Transportation Research Council; 530 Edgemont Road; Charlottesville, VA 22903; (434) 293-1997 (voice); (434) 293-1990 (fax); arkopal@virginia.edu (corresponding author)

[2]Associate Principal Research Scientist; Virginia Transportation Research Council; 530 Edgemont Road; Charlottesville, VA 22903; Phone: (434) 293-1999; Fax: (434) 293-1990; E-mail: John.Miller@VDOT.Virginia.gov

[3]L.A. Lacy Distinguished Professor of Engineering; Department of Civil Engineering, University of Virginia; 351 McCormick Road, P O Box 400742, Charlottesville, VA 22903; (434) 924-6369 (voice); (434) 977-0832 (fax); lah@virginia.edu

## Abstract

The time taken to get to the airport and the time spent in the airport terminal are two primary phases of an air passenger's journey. About 1300 departing air passengers were surveyed and individual passenger processing time at the check-in queues were collected at 5 different airports. Data shows that air passengers' pre-flight time, which is comprised of the ground travel time, processing time, and non-airport activity time, is about the same as their flight time. Passengers arrive at the terminal at suggested hours prior to departure. At the terminal, the time spent at the check-in counters as well as the security queues is not very high. It is the non-airport activity time that is a major portion of the pre-flight time. Data also suggests an inherent and high variability in the various elements of the passengers' pre-flight time.

## Introduction

Long distance travel between and origin and a destination often involves the use of multiple modes of transportation. One such example is the journey of an air passenger. It can be broadly divided in two phases – the air phase and the land phase. Passengers arrive at the originating airport and depart from the destination airport using different modes of surface transportation. This ground travel time combined with the time taken at the check-in counters and the time spent idly at the terminal diminishes the primary advantage of air travel: speed. Flight times between the cities are often the smaller portion of the total travel time, the nonflight components of the journey being longer than the flight time. Automobiles are still the predominant choice for access to the airport and the mode share of public transportation continues to be low with Reagan National Airport in Washington D.C., having the largest share (17.5%). Thus majority of the air passengers experience the same delay enroute to the airport as do other motorists since they share the roadway. Because highway congestion is temporal, an air traveler will have difficulty in predicting the ground travel time to the airport for every flight. The situation is similar at the airports. The curbside is congested, as are the terminals themselves, typically crowded by travelers, visitors, airport employees, and vendors. The access time uncertainties necessitate that the air travelers leave for the airport much earlier than may be required, thus lengthening the overall trip time by a significant amount.

There are no precise points marking where the access trip begins and where it ends. Thus a single metric capturing this access time is not generally reported. Instead, previous studies have reported either the ground travel time required to reach the airport or the level of service provided at the terminal. Our study, conducted at five airports, defines pre-flight time, which comprises of the ground travel time and the time spent by the passengers at the airport terminal, as the difference between the traveler's departure time from origin (home, work, etc.) and their scheduled boarding time. This paper identifies the elements & reports on the methods and techniques used to develop the pre-flight time.

## Data Collection

Ten pieces of information was collected at five airports on the East Coast:
1. Charlottesville-Albemarle Airport (CHO), Charlottesville, Virginia
2. Norfolk International Airport (ORF), Norfolk, Virginia
3. Baltimore/Washington International Thurgood Marshall Airport (BWI), Baltimore, Maryland
4. Richmond International Airport (RIC), Richmond, Virginia
5. Boston Logan International Airport (BOS), Boston, Massachusetts

At Baltimore, Charlottesville, Norfolk, and Richmond surveys were conducted in the airport terminal. A booth was set up and departing air passengers were approached with a request to fill out a travel questionnaire. At Boston, surveys were distributed onboard the Logan Express buses that provide transportation from three different regions to the Boson Logan International Airport. The surveys at all five locations provided the following data:
- origin in the region (zip code)
- arrival time at airport
- scheduled flight departure time
- ground travel time to airport
- perceived variability in ground travel time
- mode to access to airport
- cost of ground travel to the airport
- cost of flight ticket

At four of the five airports, passenger processing times were collected manually at the terminal. At each airport, two or three data collectors were suitably stationed with stop watches, data entry sheets, or a laptop and all collectors had a clear view of the queue and the check-in counter. No two data collectors observed the same check-in queue. Each data collector recorded the time a passenger entered the queue, the time the passenger reached the counter, and the time the passenger left the counter. To identify the variation in processing times, the process was executed at different times in a day, at different check-in queues, for two days. The two data elements collected were –

- wait-time in queue prior to check in
- service time at check-in counter

## Components of Air Passengers' Total Travel Time

Figure 1 shows the different time elements used in the study. The timeline is approximate (not to scale) and depicts a passenger's total travel time to the destination airport.

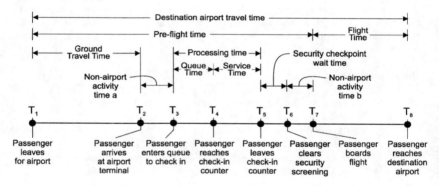

**Figure 1. Components of Air Passengers' Total Travel Time**

The following describe the terms used in Figure 1.

*Pre-flight time:* difference between the time passengers leave origin (home, work, etc.) and their scheduled boarding time

*Flight time:* time between passenger's scheduled departure time and scheduled arrival time at destination airport

*Destination airport travel time:* summation of access time and flight time

*Ground travel time:* time taken by passenger to travel from origin (home, work, etc.) to airport terminal

*Processing time:* summation of wait time in check-in queue and service time at ticketing counter:

$$\text{Processing time} = \text{Queue time} + \text{Service time}$$

*Non-airport activity time:* time spent by the passengers in traversing the terminal when not engaged in a required airport/airline procedure.

Non-airport related activity time=Non-airport activity time+ Non-airport activity

where
Non-airport activity time is the time required to walk to the check-in queue upon arriving at the airport terminal and

Non-airport related activity time is the time the time spent at the terminal after security clearance. Data pertaining to Non-airport activity time were not collected.

## Results

Tables 1 and 2 provide a summary of the findings for two sets of passengers: those arriving directly at the airport (Table 1), and those using an offsite terminal to arrive at the airport (Table 2).

**Table 1. Direct Access to Airport**

| | CHO | ORF | RIC | BWI |
|---|---|---|---|---|
| | | | | |
| Average total processing time | 4min | 12min 31sec | 5min 43sec | 7min 25sec |
| Average wait time at ticketing queues | 1min 24sec | 8min 20sec | 2min 03sec | 3min 26sec |
| Average service time at ticketing counters | 2min 36sec | 4min 11sec | 3min 40sec | 3min 59sec |
| Number of processing time observations | 323 | 340 | 346 | 423 |
| | | | | |
| | | | | |
| Number of surveys collected | 96 | 113 | 199 | 244 |
| Average ground travel time to the airport terminal | 28min | 43min | 37min | 44min |
| The predominant mode of access | auto (58%) | auto (54%) | auto (41%) | auto (34%) |
| Second most predominant mode | drop off (26%) | drop off (21%) | drop off (31%) | drop off (24%) |
| Average arrival at airport prior to scheduled departure | 1hr 4min | 2hr 22min | 2hr 10min | 2hr 2min |
| Average pre-flight time | 1hr 32min | 3hrs 03min | 2hr 45min | 2hr 47min |
| Average non-airport activity time | 1hr 2min | 2hr 08min | 2hr 03min | 1hr 57min |
| Average flight time | 3hr 12min | 2hr 52min | 2hr 40min | 2hr 32min |
| Average ticket cost | $496 | $465 | $463 | $376 |
| Average ground travel cost | $38 | $35 | $37 | $47 |
| Willing to use an offsite terminal | 53/96 | 75/113 | 144/199 | 171/244 |
| | | | | |
| | | | | |
| Average destination airport travel time | 4hr 44min | 5hr 52min | 5hr 29min | 5hr 20min |
| Average flight time *vs* destination airport travel time | 61% | 47% | 47% | 46% |
| Average ground travel time *vs* destination airport travel time | 11% | 12% | 12% | 14% |
| Average non-airport activity time *vs* destination airport travel time | 27% | 39% | 43% | 41% |
| Average ground travel time *vs* flight time | 24% | 38% | 31% | 41% |
| Average non-airport activity time *vs* flight time | 55% | 107% | 114% | 106% |

**Table 2. Indirect Access to Airport**

|  | Braintree | Woburn | Framingham |
|---|---|---|---|
| Number of surveys collected | 307 | 187 | 161 |
|  |  |  |  |
| Average ground travel time from origin to logan terminal | 28min 39sec | 23min 12sec | 24min 18sec |
| Average wait time at logan terminal | 15min 42sec | 16min 43sec | 12min 24sec |
| Average ground travel time from logan terminals to BOS | 31min 41sec | 27min 03sec | 40min 03sec |
| Average ground travel time to airport terminal | 58min 10sec | 50min 51sec | 57min 33sec |
|  |  |  |  |
| Predominant mode of access to logan terminal | drop off (55%) | drop off (47%) | drop off (54%) |
| Second most predominant mode | auto (32%) | auto (33%) | auto (36%) |
| Average flight time | 4hr 13min | 3hr 50min | 3hr 22min |
| Average time reached airport terminal prior to scheduled | 2hr 21min | 2hr 01min | 2hr 09min |
| Average ticket cost | $447 | $445 | $425 |
| Average ground travel cost | $29 | $22 | $25 |
| Require additional services at logan terminal | 221/291 | 132/181 | 109/155 |
|  |  |  |  |
| Average destination airport travel time | 7hr 33min | 6hr 31min | 6hr 08min |
| Average ground travel time vs destination airport travel time | 16% | 16% | 18% |
| Average flight time vs destination airport travel time | 52% | 54% | 51% |
| Average ground travel time vs flight time | 38% | 32% | 38% |

Figure 2 (drawn to scale), depicts the averages of the various components of the total travel time of a passenger departing from BWI.

The general trends for all airports were:

- The pre-flight time is as long as the flight time.
- The non-airport activity time is the largest portion of the pre-flight time.
- The processing time is a small portion of the pre-flight time.

Tables 1 and 2 reveal that although the average flight time varies from 2 hr 32 min (BWI) to 4 hr 13 min (Braintree) across the airports, the destination airport travel time of the passenger is much higher (varies between 4 hr 44 min (CHO) and 7 hr 33 min (Braintree)). The three elements that play a role in this finding are:

1. ground travel time
2. processing time
3. non-airport activity time

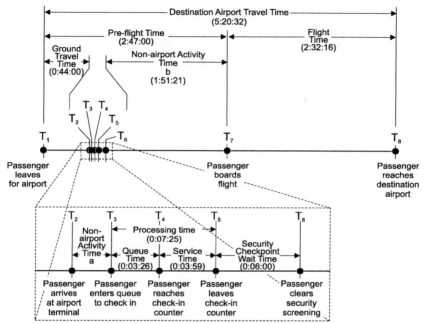

**Figure 2. Travel Timeline of Passenger Departing from BWI**

**Variability in Different Stages of Passenger Access**

The passengers arrived on average about 2 hr 6 min ahead of their scheduled departure time (which is within the time frame suggested by many airlines), but the variation in times was high. Some passengers arrived as late as 10 min prior to departure, and some arrived as early as 9 hr prior to departure.

Although the average *arrival time at airport prior to scheduled departure* is consistent with the advance arrival times that airlines require of their passengers, we observe a high coefficient of variation across all airports (Table 3). Reasons for this variation can be partly attributed the passengers' perceived variability of ground travel time to the airport and partly to their perceived variability in the airport's processing times. Table 3 could also suggest that passengers are more certain about their ground travel time while accessing the airport indirectly (i.e., using an offsite facility such as the Logan Express terminals) than while accessing the airport directly (higher coefficient of variation). One reason might be that the Logan Express buses use HOV lanes, thus reducing passenger uncertainty.

Table 4 shows the ground travel times of passengers traveling from a particular zip code to a particular airport. It is observed that the variability in ground travel times is lower than the ones observed in Table 3.

**Table 3. Arrival Time at Airport Prior to Scheduled Departure**

|  | Arrival at airport prior to scheduled departure | | |
|---|---|---|---|
|  | Mean | Median | Coefficient of variation |
| CHO | 1hr 4min | 1hr | 46% |
| ORF | 2hr 22min | 2hr | 66% |
| RIC | 2hr 10min | 1hr 47min | 65% |
| BWI | 2hr 2min | 1hr 45min | 54% |
| BOS - Braintree terminal | 2hr 21min | 2hr 9min | 50% |
| BOS - Woburn terminal | 2hr 1min | 2hr | 42% |
| BOS - Framingham terminal | 2hr 9min | 2hr | 47% |

**Table 4. Ground Travel Time Between Origin and Airport**

|  |  | Ground Travel Time to Airport | |
|---|---|---|---|
| Airports | Top Passenger Generating Zip Code | Mean | Coefficient of Variation |
| CHO | 22901 | 19 min | 52% |
| ORF | 23454 | 40 min | 23% |
| RIC | 23112 | 37 min | 17% |
| BWI | 21401 | 28 min | 20% |

Once at the airport terminal, the passengers must get in the check-in queues (unless they have an e-ticket and a boarding pass and no bags to check in) and then the security queues. Only data pertaining to check-in queues were collected. Data was obtained from the Transportation Security Agency (TSA) website which provides historical average and maximum wait times of passengers at security screening queues (not the individual wait times of passengers). Based on this data collected between April 7, 2007 & May 10, 2007, Table 5 shows the average security screening times. It is observed that the security wait times on an average is 4% of the passengers' time spent at the airport terminal. Also, the processing time at ticketing counters, although a very small proportion of the passengers' pre-flight time (3% as shown in Figure 2), is highly variable (shown in table 6).

**Table 5. Security Checkpoint Wait Time**

|  | Security checkpoint wait time | | |
|---|---|---|---|
|  | Average | Median | Maximum |
| CHO | 4 min | 4 min | 10 min |
| ORF | 5 min | 4 min | 21 min |
| RIC | 2 min | 2 min | 12 min |
| BWI | 6 min | 5 min | 28 min |
| BOS | 4 min | 3 min | 36 min |

**Table 6. Processing Time**

|  | Total processing time | | |
|---|---|---|---|
|  | Mean | Median | Coefficient of variation |
| CHO | 4min | 3min 5sec | 74% |
| ORF | 12min 31sec | 8min 38sec | 104% |
| RIC | 5min 43sec | 3min 50sec | 90% |
| BWI | 7min 25sec | 5min 52sec | 71% |

Once past the security check point, the passengers indulge in non-airport activities prior to boarding the flight. This non-airport activity time is not only comparable to the flight time but also is a major portion of the pre-flight time. An average passenger arrives at CHO 1hr 4min prior to the scheduled departure (shown in Table 3), spends 4 min in the check-in process (shown in Table 6), and spends an additional 4 min clearing the security checkpoint (shown in Table 5). Hence it is seen that the average passenger's non-airport activity time at CHO is 88% of his or her time spent at the airport terminal.

The two data elements used to calculate the non-airport activity time, i.e., processing time and arrival at the airport prior to scheduled departure, were collected using different methods. Thus, a given passenger's processing time cannot be matched with his or her arrival at the airport prior to departure. Hence the non-airport activity time was determined by simulating 50 combinations of linking the two data elements. For 95% of the simulations, the range of non-airport activity time is shown in Table 7.

### Table 7. Non-Airport Activity Time

|  | Non-airport activity time | | |
|  | Mean | Median | Coefficient of variation |
|---|---|---|---|
| CHO | 1hr 1min | 57min 17sec | 49% |
| ORF | 2hr 08min | 1hr 50min | 75% |
| RIC | 2hr 02min | 1hr 37min | 68% |
| BWI | 1hr 57min | 1hr 42min | 57% |

### Discussion

The high coefficient of variation in the arrival at airport prior to scheduled departure (Table 3), the processing time (Table 6), and the non-airport activity time (Table 7) led the researchers to investigate if the variability was inherent or attributable to outliers in the data.

A box plot diagram is a good indicator of dispersion in data sets. Figure 3 is a box plot diagram of the arrival at airport prior to scheduled departure data for all survey locations. The height of the box (i.e., the interquartile range, Q3-Q1) indicates the variability in the data sets. It can be seen that the CHO data have less variation than the ORF data. The location of the median line helps indicate how symmetrical the data are. If it is far from the center of the box, the distribution is skewed. The diagram also helps identify the outliers in the data.

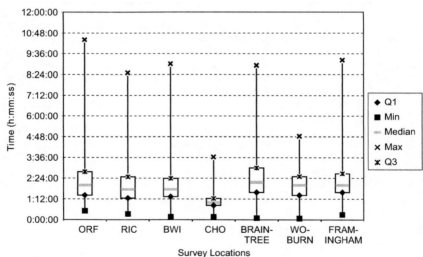

**Figure 3. Original Data Set for Arrival at Airport Prior to Scheduled Departure**

As shown in Table 3, for passengers departing from ORF, the *arrival at airport prior to schedule departure* data have a mean of 2 hr 22 min and a standard deviation of 1 hr 34 min (which gives a coefficient of variation of 66%); the median is 2 hr. Eliminating the values smaller than 95% of the population and the values larger than 95% of the population decreased the coefficient of variation to 42%. A 95% confidence interval for the population mean shows that the range of the mean for the arrival at airport prior to scheduled departure lies between 1 hr 49 min and 2 hr 20 min. A hypothesis test conducted for the mean processing time of 2 hr could not be rejected at the 5% significance level. Thus, the mean and median are not significantly different.

The removal of outliers from the data did lower the variability, but it was still high (as shown in Table 8). This led the researchers to the conclusion that there is a high inherent variability in various phases of passenger access to the airport.

**Table 8. Change in Coefficient of Variation for Arrival at Airport prior to Scheduled Departure**

|  | Characteristic | Arrival at airport prior to scheduled departure | |
|---|---|---|---|
|  |  | Complete Data Set | Data set with no outliers |
| CHO | Rural, no offsite facility | 46% | 26% |
| ORF | Urban, no offsite facility | 66% | 42% |
| RIC | Urban, no offsite facility | 65% | 44% |
| BWI | Urban, no offsite facility | 54% | 38% |
| BOS - Braintree terminal | Urban, offsite facility | 50% | 41% |
| BOS - Woburn terminal | Urban, offsite facility | 42% | 39% |
| BOS - Framingham terminal | Urban, offsite facility | 47% | 38% |

## Study Limitations

1. For our study, there is not sufficient information available regarding the exact address of the air passenger's origin within the zip code. To the extent that origins within a single zip code exhibit different travel times to the airport, the use of a zip code rather than an exact address may mask some of the travel time variability.
2. Because most of the travelers use an auto to access the airport, the analysis of ground travel time and its variation contains limited information regarding other modes.
3. Variability in the security checkpoint wait times could not be developed due to the unavailability of individual screening times.

## Conclusions

The quality of the air traveler's experience is affected by both (a) the length of the pre-flight components of the journey and (b) the uncertainty associated with how long these pre-flight components will take. Subsequently, this study draws the following three conclusions.

1. Pre-flight time is a large proportion of the destination airport travel time

Compared to other modes, the principal advantage of air travel is speed. With average pre-flight times as high as 3hr 3min, large pre-flight times are especially a disadvantage to the travelers who have short flight times. Such high pre-flight times could act as a deterrent for these travelers, who might instead decide to travel to their destination using alternate modes of transport.

2. Ground travel time and non-airport activity time affect an air passenger's pre-flight time

Passengers arrive at the airport from various origins in the region, which are as far as 140 miles from the airport. Thus passengers not only have high ground travel times based on their origins in the region, but also face high variability in these times. Similarly, it was observed that the passengers spend a large proportion of their pre-flight time at the airport terminal. During their time at the terminal, the passengers spent only a small fraction of their time engaged in required airline/airport procedures such as ticketing and security clearance. Non-airline/airport related activities accounted for the majority of their time. (Some of the non-airport activity time may be spent shopping, which may benefit both, the traveler and the airport. However, the rest of the non-airport activity time may not be of benefit to either, such as time spent waiting at the boarding lounge).

3. Processing time shows high variability

Processing time at the check-in counters, although a small proportion of the passengers' pre-flight time, shows high variability. Consequently passengers are highly uncertain of their arrival time at the airport prior to their scheduled departure. Thus the variability in processing time, rather than the processing time itself, is one of the reasons airlines recommend their passengers arrive hours prior to their scheduled departure.

**Recommendations for further research**

An improvement to the air traveler's destination airport travel time could be achieved by reducing the variability in the processing times. This in-turn will reduce the length of the non-airport activity time, further shortening the destination airport travel time. If the uncertainty were to be minimized, airlines could recommend their passengers to arrive at the airport terminal at a time which is closer to their scheduled boarding time. By doing so, the non-productive segment of the non-airport activity time would be reduced, causing a significant reduction in the air passenger's pre-flight time.

**References**

Mahmassani, H.S., Slaughter, K., Chebli, H., and McNerney, A. (2001) Domestic and International Best Practice Case Studies, Center of Transportation Research, University of Texas at Austin.

Traffic Congestion and Reliability: Linking Solutions to Problems. (2004) Prepared by Cambridge Systematics Inc., with Texas Transportation Institute for Federal Highway Administration, Cambridge, Massachusetts http://www.ops.fhwa.dot.gov/congestion_report/chapter1.htm. (September 26, 2006)

Tunasar, C., Bender, G., and Young, H. (1998) Modeling Curbside Vehicular Traffic at Airports, In Proceedings of the Winter Simulation Conference, SABRE Group, Southlake, Texas.

Goswami, A.K., Miller, J.S., Hoel, L.A. (2005) Assessing the Feasibility of Passenger Offsite Airport Terminals, 9th Air Transport Research Society (ATRS), Rio de Janeiro, Brazil.

# How Much Vehicular Traffic Do Airports Generate?

Srinivas S. Pulugurtha[1] and Timothy McCall[2]

[1]Assistant Professor of Civil Engineering, Assistant Director of Center for Transportation Policy Studies, The University of North Carolina at Charlotte, 9201 University City Boulevard, Charlotte, NC 28223-0001, USA; PH: (704) 687-6660, FAX: (704) 687-6953, email: sspulugu@uncc.edu
[2]Undergraduate Student of Civil Engineering, The University of North Carolina at Charlotte, 9201 University City Boulevard, Charlotte, NC 28223-0001, USA; PH: (704) 687-2304, FAX: (704) 687-6953, email: tcmccall@uncc.edu

## Abstract

Airports serve different regions and different economies. Some airports serve as an origin and destination (O&D) point to point system while others serve as a hub and spoke system. The type of airport and modes of transportation available to the airport cause great variations in traffic volumes on roads in the vicinity of the airport. In particular, O&D point to point airports have a profound effect on the local road infrastructure. Research documents very few studies to estimate the number of trips generated by airports. In the most current volume of the Institute of Transportation Engineers (ITE) Trip Generation Manual, only 3 commercial airports were analyzed to determine comparisons of employees, passengers, or airport demand versus trip generation. Anecdotal evidence indicates that estimates of number of trips generated by airports based on these studies are very different from the observed data. The focus of this paper is to research and study the relationships between airport type, airport demand, and traffic volumes. Data such as the type of airport, airport demand (total number of enplaning and deplaning passengers), total enplaned and deplaned cargo in tons, number of airline operations, number of employees, metropolitan area population, presence of large scale transit systems, the percent of connecting passengers, and traffic volumes along a major access road connecting arrival/departure gates at selected airports in the United States were collected and analyzed to study the relationships and develop equations to estimate the number of trips generated by the airports. The results from this research are expected to help planners and engineers better identify required infrastructure for improved traffic operations at existing and new airports.

## Introduction

Airports serve different regions for a variety of reasons. Some regions rely heavily on airports as a source of revenue for freight operations. Others rely less on their airports, yet their airports serve as hubs for major airlines. Vacation areas rely on airports, in particular, if their location is relatively remote with limited access.

Providing accessibility to the airport within the region or city through appropriate road infrastructure has been vital for successful operations at many airports.

How are vehicle trips to airports measured? Currently, there are only trip generations for airports based on the number of employees and yearly passenger figures (ITE 2003). The equations for commercial service airports were established based on two California studies in 1975 and one in 1983. Trends or relationship between trip generation rate and number of "origin and destination" (O&D) were also observed (Ruhl and Trnavskis, 1998). In another study, De Neufville and Odoni (2003) state that an airport that serves 10 million passengers per year generates about 40,000 trips per day. Anecdotal evidence shows that when incorporating these figures into estimated trip generations, the value obtained was different from actual traffic volume counts for most airports.

The vehicular traffic generated by airports depends on the total number of enplaning and deplaning passengers. However, several airports are used by major airlines as a hub. These airports serve one or more major airlines and support both large and regional aircraft. The number of connecting (transferring) passengers at such airports is very high. Examples of such airports include Charlotte-Douglas International Airport, Cincinnati / Northern Kentucky Airport, Minneapolis – St. Paul International Airport and Denver International Airport where over 40 percent of passengers transfer to other airports. This can ultimately put less strain on the surrounding infrastructure since most passengers will be connecting to other cities. Using the total number of enplaning and deplaning passengers without considering the number or percent of connecting passengers may lead to overestimates in such cases.

Variables such as enplaned and deplaned cargo in tons (which is an indicator of regional economy) and the number of airline operations may have a bearing on vehicular traffic generated by airport. Intuitively, it can be stated that the number of operations will be large in number if the total number of passengers or cargo is higher. The numbers of employees who work at airport also contribute to vehicular traffic. Such employees include those who work in administration department, security department, those who work for airlines, rental car companies, food concession, shops, restaurants, and parking. Activity at an airport may be directly proportional to the socio-economic and demographic characteristics of the metropolitan area (example, population). Passengers and employees may use public transportation systems such as subway or rail if such a service exists.

Thus, in general, trip generations vary in relation to airport function as either an O&D point to point or a hub airport (as a function of percent of connecting passengers), the number of passengers, cargo handled, number of airline operations, metropolitan area population, and if the airport has a subway or rail link. The question is how these variables are related to vehicular traffic generated by airports. In this paper, data pertaining to the discussed variables were collected and analyzed for twenty airports in the United States. The results from linear regression analyses conducted provide more realistic insights to estimate vehicular traffic generated by airports. Such estimates will allow planners and engineers to effectively predict the affects of traffic due to airport expansion, new airport construction, and airport operations in a given location or city.

**Data Collection**

As stated before, data was collected for twenty airports in the United States. Attempts were made to select airports so that they are geographically distributed throughout the United States. The following data was collected for each selected airport.

1) Type of airport
2) Total number of enplaning and deplaning passengers
3) Total enplaned and deplaned cargo in tons
4) Number of airline operations
5) Number of employees working at the airport
6) Population of the metropolitan area
7) Percent of connecting passengers
8) If the airport has a subway or rail link
9) Major access road and traffic volume along this road near the airport

In this report, airports were classified as large hub airports, medium hub airports, or small or non-hub airports (FAA 1997). Data pertaining to the, presence of a subway or rail link was obtained from the "Highlights of the 2003 General Information Survey" published by Airports Council International (ACI 2007). The total number of enplaning and deplaning passengers, total enplaned and deplaned cargo, and number of airline operations in 2005 was obtained from traffic statistics (or facts) published by each individual airport and the Airport Council International. The number of employees working at the airport in 2005 (most cases where data was available) and the percent of connecting passengers was obtained from facts and publications available on the Internet.

The main access road and traffic volume along the main access road near the airport in 2005 was obtained from road maps and annual traffic reports published by the State Department of Transportation or other responsible transportation agency in which the airport is located. Table 1 shows a summary of data collected for each selected airport.

**Regression Analysis and Models**

An off-the-shelf statistical software program, MINITAB ® (Minitab 2003) was used to conduct the linear regression analysis. As a first step, linear regression analysis was conducted to study the relation between each selected variable (considering it as an independent variable) and the annual traffic volume along the main access road (dependent variable) near the airport. A summary of results obtained for each selected variable are shown in Table 2.

Results indicate that the coefficient for each variable in each model is positive. The coefficient for airport type indicates that traffic volume along access road near airport is lowest at small or non-hub airports whereas it is the highest at large hub airports followed by medium hub airports. This is primarily due to the high number of enplaning and deplaning passengers at large hub airports. Note that according to Federal Aviation Administration (FAA 1997), an airport with 1 percent or more of total U.S. passengers qualifies as a large hub airport whereas an airport with 0.25 to

# Table 1. Summary of data collected for each selected airport.

| Airport Name | Airport Code | Airport Type[1] | | | Passengers Total | Cargo (tons) | Airline Operations | # Employees[2] | Population | % Connecting | Subway or Rail Link[3] | Main Access Road | Access Road Annual Traffic |
|---|---|---|---|---|---|---|---|---|---|---|---|---|---|
| | | LH | MH | SH | | | | | | | | | |
| Las Vegas McCarran Intl | LAS | 1 | 0 | 0 | 43,989,912 | 100,483 | 605,046 | 12,000 | 1,710,551 | 8 | 0 | Paradise | 13,067,000 |
| Denver Intl | DEN | 1 | 0 | 0 | 43,387,513 | 309,848 | 560,669 | 30,000 | 2,359,994 | 44 | 0 | Pena Blvd | 12,410,000 |
| Phoenix Sky Harbor Intl | PHX | 1 | 0 | 0 | 41,204,071 | 302,197 | 555,256 | 31,437 | 3,865,077 | 31 | 0 | Sr-153 | 21,316,000 |
| Minneapolis-St Paul Intl | MSP | 1 | 0 | 0 | 36,678,868 | 282,422 | 532,240 | 5,273 | 3,142,779 | 54 | 0 | Glumack Dr | 2,920,000 |
| Detroit Intl | DTW | 1 | 0 | 0 | 36,402,710 | 220,721 | 521,899 | 18,000 | 4,488,335 | 55 | 0 | John D Dingell Drive | 10,090,790 |
| Miami Intl | MIA | 1 | 0 | 0 | 31,008,453 | 1,754,633 | 381,610 | 11,357 | 5,422,200 | 38 | 0 | Airport Expressway | 35,952,500 |
| Seattle-Tacoma Intl | SEA | 1 | 0 | 0 | 29,289,026 | 338,591 | 341,762 | 6,939 | 3,203,314 | 26 | 0 | International Blvd | 19,346,460 |
| Charlotte-Douglas Intl | CLT | 1 | 0 | 0 | 28,206,052 | 158,899 | 521,878 | 18,524 | 1,521,278 | 75 | 0 | Josh Birmingham Parkway | 7,300,000 |
| Boston Logan Intl | BOS | 1 | 0 | 0 | 27,087,905 | 356,120 | 409,066 | 12,000 | 4,411,835 | 10 | 1 | I-90 West of Exit 25-24 | 15,111,000 |
| Cincinnati/Northern KY Intl | CVG | 1 | 0 | 0 | 22,778,785 | 251,602 | 496,364 | 6,000 | 2,070,441 | 65 | 0 | KY 212 | 11,011,320 |
| Washington National | DCA | 1 | 0 | 0 | 17,847,884 | 3,969 | 276,056 | 9,155 | 5,214,666 | 10 | 1 | SR 233 | 6,570,000 |
| Portland Intl | PDX | 0 | 1 | 0 | 13,879,701 | 261,473 | 263,253 | 18,712 | 2,095,861 | 17 | 0 | Airport Way | 19,793,950 |
| Memphis Intl | MEM | 0 | 1 | 0 | 10,963,000 | 3,598,501 | 392,360 | 31,000 | 1,260,905 | 61 | 0 | Winchester Road | 6,450,645 |
| Raleigh-Durham Intl | RDU | 0 | 1 | 0 | 9,303,904 | 109,768 | 248,511 | 2,300 | 1,405,868 | 2 | 0 | Terminal Blvd/Aviation Pkwy | 11,315,000 |
| Ontario Intl | ONT | 0 | 1 | 0 | 7,213,528 | 521,859 | 143,249 | 6,000 | 12,923,457 | 2 | 0 | Archibald Ave | 5,374,625 |
| Reno-Tahoe Intl | RNO | 0 | 1 | 0 | 5,200,000 | 50,234 | 140,851 | 2,600 | 393,946 | 0 | 0 | Plumb Lane | 6,716,000 |
| Omaha Eppley Airfield | OMA | 0 | 1 | 0 | 4,192,046 | 87,249 | 144,150 | 1,800 | 813,170 | 0 | 0 | Abbott Dr | 7,445,270 |
| Spokane Intl | GEG | 0 | 1 | 0 | 3,197,440 | 52,263 | 97,923 | 1,312 | 440,706 | 0 | 0 | W Airport Dr | 365,000 |
| Piedmont Triad Intl | GSO | 0 | 0 | 1 | 2,600,924 | 73,552 | 129,419 | 4,861 | 674,500 | 0 | 0 | Joseph M Bryan Blvd | 1,629,360 |
| Greenville-Spartanburg | GSP | 0 | 0 | 1 | 1,575,117 | 22,000 | 25,185 | 758 | 858,060 | 0 | 0 | Airport Terminal Road | 2,555,000 |

[1] LH, MH, and SH indicate large hub, medium hub, and small hub, respectively.
[2] # employees working at the airport.
[3] 1 indicates subway or rail link exists; 0 otherwise.

0.99 percent of total U.S. passengers qualifies as a medium hub airport. Small and non-hub airports enplane 0.05 to 0.249 percent and less than 0.05 percent of total U.S. passengers, respectively. The T-statistic is greater than 2 and the P-value is less than 0.05 for variables "large hub" and "medium hub". Though the T-statistic is less than 2.0 and the P-value

**Table 2. Linear regression analysis between each variable and traffic volume – summary.**

| Variable | | Coefficient | T-Statistic | P-value | F-statistic | P-value |
|---|---|---|---|---|---|---|
| Airport Type | Large Hub | 14,099,552.00 | 5.99 | 0.000 | | |
| | Medium Hub | 8,208,641.00 | 2.78 | 0.013 | 14.60 | 0.00 |
| | Small Hub | 2,092,180.00 | 0.38 | 0.709 | | |
| Passengers | | 0.44 | 6.35 | 0.000 | 40.36 | 0.00 |
| Cargo | | 7.46 | 2.56 | 0.019 | 6.55 | 0.02 |
| Airline Operations | | 28.81 | 6.07 | 0.000 | 36.84 | 0.00 |
| # Employees | | 669.70 | 4.84 | 0.000 | 23.40 | 0.00 |
| Population | | 2.25 | 3.88 | 0.001 | 15.03 | 0.00 |
| Percent of Connecting Passengers | | 244,586.00 | 3.60 | 0.002 | 12.95 | 0.00 |

is greater than 0.05 for small hub, the F-statistic is greater than 4 and the P-value is less than 0.05 for the overall model indicating that traffic volume depends on airport types such as large hub, medium hub, and small hub.

The T-statistic is greater than 2 and the P-value is less than 0.05 when models were developed using the total number of passengers, cargo in tons, the annual airline operations, the number of employees working at the airport, population, and the percent of connecting passengers as an independent variable. The F-statistic is greater than 4 and P-value is less than 0.05 for all the models indicating that traffic volume depends on each of these variables when modeled separately. The F-statistic and T-statistic are highest when the number passengers was used followed by annual airline operations and the number of employees working at the airport. Note that the coefficient for the percent of connecting passengers is positive which intuitively do not make sense.

In summary, the analysis indicate that large hub and medium hub airport type, the total number of enplaning and deplaning passengers, number of airline operations (which include cargo operations), the number of employees working at the airport, and population may have a bearing on the vehicular traffic generated by airports. However, these variables may be related to each other.

As populated areas generate more air travel demand (passengers) and FAA's definition of airport type depends on passenger demand, linear regression model was developed by using annual traffic volume along main access road near the airport as a dependent variable and the total number of enplaning/deplaning passengers, the number of employees working at the airport, and the percent of connecting passengers as independent variables. A summary of the results obtained are shown in Table 3. From the table, it can be said that each passenger generates 0.4 vehicle trips whereas each employee generates 225 vehicle trips during a year. The T-statistic is greater than 2 and the P-value is lower than 0.05 only for passengers. However, the F-statistic is greater than 4 and the P-value is lower than 0.05 for the overall model.

As vehicular traffic generated depends on cargo in addition to passengers, a linear regression model was developed by using annual traffic volume along main access road near the airport as a dependent variable and annual airline operations, the number of employees working at the airport, and the percent of connecting passengers as independent variables. Passengers were not considered in this case as airline operations include both passengers and cargo operations. A summary of the results obtained are shown in Table 4. From the table, it can be said that each airline operation generates 32 vehicle trips whereas each employee generates 187 vehicle trips during a year. The T-statistic is greater than 2 and the P-value is lower than 0.05 for only airline operations. The F-statistic is greater than 4 and the P-value is lower than 0.05 for the overall model.

**Table 3. Linear regression analysis results summary between passengers, employees, percent of connecting passengers, and traffic volume.**

| Variable | Coefficient | T-Statistic | P-value | |
|---|---|---|---|---|
| Passengers | 0.40 | 2.81 | 0.012 | F-statistic = 13.10 |
| Employees | 225 | 0.96 | 0.351 | P-value = 0.000 |
| Percent of Connecting Passengers | -61,727.00 | -0.66 | 0.516 | |

**Table 4. Linear regression analysis results summary between airline operations, employees, percent of connecting passengers, and traffic volume.**

| Variable | Coefficient | T-Statistic | P-value | |
|---|---|---|---|---|
| Airline Operations | 31.56 | 2.65 | 0.017 | F-statistic = 12.45 |
| Employees | 187.2 | 0.75 | 0.464 | P-value = 0.000 |
| Percent of Connecting Passengers | -114,377.00 | -1.08 | 0.293 | |

Results from the models shown in Table 3 and Table 4 indicate that annual traffic volume along main access road near the airport decreases as the percent of connecting passengers increase. However, the T-statistic and P-value indicate that the relation may not be significant enough to explain the relation. The models based on total number of passengers and airline operations appear to be the best and can be considered to estimate vehicular traffic along main access roads near airports.

## Conclusions

This paper presents a linear regression analysis to identify variables to model and estimate vehicular traffic generated by airports. Results shows that variables such as airport type (large hub and medium hub), total number of enplaning and deplaning passengers, cargo in tons, the number of airline operations, population, the number of employees working at the airport, and the percent of connecting passengers may all have a bearing on the vehicular traffic generated by airports when modeled individually. Based on the models developed in this study, vehicular traffic generated along main access road near airports can be better estimated using the total number of passengers or number of airline operations. However, it is felt that the effect of the percent of connecting passengers on vehicular traffic generated near airports need to be further studied by 1) considering a larger sample size, and 2) exploring the possibility of a non-linear relationship.

Only one main access road for each airport was considered in this paper. Major airports may have multiple main access roads to reduce congestion and allow smooth flow of traffic in the vicinity of airport. The presence of such access roads and vehicular traffic on those roads have to be included in developing models to estimate vehicular traffic generated by airports.

Also, main access roads may be used by local users (residents) in addition to those traveling to the airport. Such factors should also be considered so as not to overestimate vehicular traffic generated by airports.

## References

Airports Council International - ACI (2007) Highlights of the 2003 General Information Survey. http://www.aci-na.org/asp/stats.asp?page=93, Webpage Accessed on February 28, 2007.

De Neufville, R. and A. Odoni (2003) Airport Systems: Planning, Design, and Management. McGraw Hill, New York.

FAA (1997) Terminal Area Forecasts. Final Report (FAA-APO-97-7), United States Department of Transportation, Federal Aviation Administration.

ITE (2003) Trip Generation. Institute of Transportation Engineers, 7th Edition, Washington, DC.

Minitab (2003). Minitab Release 14 for Windows. Minitab Inc.

Ruhl, T. A. and B. Trnavskis (1998) Airport Trip Generation. ITE Journal, Vol. 68(5), 90-93.

# Information Propagation for informing Special Population Subgroups about New Ground Transportation Services at Airports

## Basil C. Stephanis[1] and Dimitris J. Dimitriou[2]

[1] ASCE member, Department of Transportation Planning, School of Civil Engineering, Democritus University of Thrace, 12, Vas. Sofias, 67100, Xanthi, Greece; PH +30 210 3724900; FAX: +30 210 3724904, email: stefanis.v@gr.selonda.com

[2] Corresponding author, Department of Civil Works, Sector of Transportation, Technical College of Athens, 338, Patission str, 11141, Athens, Greece PH +30 210 2114450; FAX +30 210 2286817 email: dimitriou@otenet.gr.

## Abstract

Implementing new transportation services for servicing only a special population subgroup presents a severe problem on transport operation process because one has to separate them from the common users and provide them specific transit services under the quality and cost restrictions. A significant counterpart of this problem is the definition of the time lag required, between the initial marketing campaign and the starting time of the service itself. Informing the potential users by mass media may lead to misuse of monies with doubtful results especially in the case of handicapped people. This population group is, on one hand scattered within the population, but on the other hand has special communication habits and the use of "word of mouth" constitutes an efficient information process. The viability of a project dealing with services for these special population subgroups mainly depends upon the number of people aware of a particular service, and the required time for the information dissemination within the population studied. This paper discusses a modeling approach based on Markov stochastic process to simulate the person-to-person information dissemination process. Application of this model focuses on the elderly and handicapped subpopulation group and takes into account the communication behaviors and habits of three separate subgroups to estimate the required time to be informed, prior to the commencement of the service.

## Introduction

An essential element in the operation planning process of new airport ground transportation services is the dissemination of new information to the impacted population subgroup. Techniques used are amply described in literature dealing with marketing, advertising and promotion of such services. A significant counterpart of this problem is the definition of the 'time lag' required between the initial marketing campaign and the starting time of the service itself. Informing the potential users by mass media may lead to misuse of resources with doubtful results, especially in the case of elderly and handicapped people.

It has been established that the service advertising cost increases and advertising effectiveness decreases with the reduction of population income. The elderly and handicapped population group studied typically has lower incomes. While this population group is scattered within the targeted population, it has certain special characteristics related to information dissemination processes that are based largely on person-to-person contact. Therefore, the use of mass media for transmitting service-related information presents a partially inefficient method.

On the other hand, a commercial announcement or marketing the service would be relatively ineffective, because it is addressed to the elderly and handicapped organizations that constitute a small fraction of the total target population. The viability of a project dealing with services to this special population group would largely depend upon the number of people becoming aware of the new service before its actual commencement. It is not necessary for the service operator to begin operation after the entire population are informed about it, but in fact it may be possible with this procedure, to allow for the project to begin as soon as the critical mass (number of people who will render the service viable) of potential users has been reached. In this manner the news will reach the entire subpopulation through the dynamics of information dissemination and diffusion

The objective of this paper is to estimate the time lag required, between the initial marketing campaign and the starting time of the service. The formulation framework that is presented here is based on a stochastic Markov approach where the stages and the parameters of the person-to-person information dissemination are analyzed and the informed and uninformed proportions are estimated. These proportions are dependent on the probability of transmitting, the communication habits and the time it takes for the news to be propagated. The initiated application involves ground transportation services from/to airports that are directed to the elderly and handicapped community.

The content of the paper is organized in three main sections. First, the literature dealing with the promotional strategy is presented. Second, the modeling framework is described and a numerical application and the results are analyzed. Finally, the conclusions reached are stated.

## Literature Review

In today's environment of increasing pressure for marketing productivity and brand profitability, the companies are seeking direction in their advertising and promotion allocation decisions in order to gain market share. Mela et al (1997); Anselmi (2000); and Sethuraman and Neslin(2000), have presented an understanding of this complex allocation decision by identifying factors that may influence the frequently opposing strategies of advertising and promotion, suggesting consumer decision-making outcomes and offering prescriptions that lead to allocation plan optimization. Several marketing researchers have empirically and analytically addressed both the depth and the frequency of promotion decisions by using data analysis tools or presenting survey results. In both streams of entrepreneurship and marketing literature, it has been recognized that the information dissemination process has significant effect on marketing decision-making as well analyzed by Kurataa and Liub (2006); Cheng and Sethi (1999); Bronnenberg (1998); Sogomonian and Tang (1993).

Public relations activities, such as press releases and charity sponsorships are low-key communication activities that a retailer can undertake within his or her local community. Fama and Yangb (2005) claim that retailers tend to place greater emphasis on below-the-line promotions (direct marketing and sales promotion) than above-the-line promotions (traditional mass media such as television, radio and print). Usually, the effect of increases or decreases in advertising is measured with a statistical measure called 'advertising elasticity'. The complicated calculations of advertising elasticity have been made with hundreds of brands and in several industry markets, as presented by Bauman et. al. (2001); Jones (2004); and McClure and Kumcu (2006).

On the other hand, over the past few years, policy makers have shown an increasing interest in social inclusion issues and reintegration protection policies for people with disabilities. However, the lack of homogeneous and specific statistical information are difficult to evaluate just how strong an impact of a relevant protection policy is having on the improvement of employment and quality of life of disabled people. In Europe, as Eurostat (2001) and Pascual and Cantarerol (2006) have recognized that disabled people typically have a relatively low educational and income level compared with non-disabled people. However, this can be questioned if used in general sense.

Memory is one of the most highly researched areas in elderly and handicapped population. A significant research about destination advertising that focused on age and format effects on memory are presented by MacKay and Smith (2006) and Smith and MacKay (2001). Today, the literature provides an incomplete picture of how the individual's memory system functions when everyday stimuli are cognitively processed. This would include messages received via mass media (advertising) or verbal media (aural and/or written and/or visual in nature).

One of the most significant protection policies available to disabled people is to provide them with specific transit services in order to upgrade their mobility. The most known application for the Handicapped Person Transportation (HPT) problem arises in the door-to-door transportation services, where many airports in Europe have applied relevant transit services from/to airport. These people are transported either in groups or individually between specified origins or destinations, and the HPT is often formulated taking into account two requests: an outbound request from home to airport, and an inbound request for airport to a specific destination. Stephanis and Dimitriou (2006), Rekieka et al.(2006) and Palmer (2004) have been engaged in research on the HPT or similar problems taking into account that the operating expenses for those mandated services have increased as demand for this has expanded.

**Modeling approach**

*Notations*

According to Kentall (1956); Daley and Kendall (1956); Bailey (1960); Bhat and Miller (2002); and Hillier and Lieberman (2005) formulating a problem as a Markov chain process should be necessary to determine a finite number of states and the stationary transitional probabilities. Stochastic processes describe the behavior of a system or procedure over some period of time. A stochastic process is defined to be an indexed collection of random variables $\{Xt\}$, where the index t runs through the given T. Thus, for $t= 0, 1, 2, ..., tk$ the random variable $Xt$ takes on the values $\{Xt\} = \{X0, X1, X2, ...Xtk\}$ provides a mathematical representation of how the status of the behavior in a system or procedure evolves over time. The stochastic process $\{Xt\}$ is said to have the Markovian property (Markov chain) if the conditional probability of any future "event" is defined by any past "events" and the present state is independent of the past event but depends only upon the present state.

The handiest form for presenting the conditional probability $p_{ij}^{(k)}$ for all the $k$-step (the probability that the system will be in the state $j$ after exactly $k$ steps e.g. time units) is given in following matrix form:

**Table 1.** Matrix forms for present the transition probabilities.

$$P^{(k)} = \begin{array}{c|cccc}
\text{State} & 0 & 1 & ... & m \\
\hline
0 & p_{00}^{(k)} & p_{01}^{(k)} & \cdots & p_{0m}^{(k)} \\
1 & p_{10}^{(k)} & p_{11}^{(k)} & \cdots & p_{1m}^{(k)} \\
... & ... & ... & ... & ... \\
m & p_{m0}^{(k)} & p_{m1}^{(k)} & \cdots & p_{mm}^{(k)}
\end{array}$$

$p_{ij} = P\{Xt = j \, / \, Xt = i\}$ is the first step and

$p_{ij}^{(k)} = P\{Xt = j \, / \, Xt = i\}$ is $k$-step of conditional probability

If the transitional (conditional) probabilities $P\{X_{t+1} = j/X_t = i\}$ for each step (stratum) and each $i,j$ is $P\{X_{t+1} = j/X_t = i\} = P\{X_1 = j/X_0 = i\}$, for all $t=1,2,...t_k$ implies that the transition probabilities do not change over time. The conditional probability $p_{ij}^{(k)}$ of any future "event" must be positive number and since the process must make a transition into some state, it must satisfy the properties:

$p_{ij}^{(k)} \neq 0$ for all $i$ and $j$ - where $k = 0,1,2,...$

and

$$\sum_{j=0}^{m} p_{ij}^{(k)} = 1 \text{ for all } i\,j \text{ - where } k = 0,1,2...$$

## Propagation Assumptions

The person-to-person conditions for the information dissemination process depend on group element characteristics. The corresponding frame structures for all possible contact conditions are:

(a) One–to-one contract: a frame unit of exclusive groups element corresponds to only one population element
(b) One-to-many: one frame unit corresponds to more than one population element
(c) Many-to-one: several frame units correspond to one population element. Here, it is possible that a population element could appear multiple times
(d) Simple many-to-many: several frame units correspond at least to one population element or reversely
(e) Complex many-to-many: Including all frames that cannot be classified into one of the above described structures.

It is assumed that the studied population group size $S$ could be recognized, and it involves the following main subgroups:

(a) the population to be informed $N$ and
(b) "the seed" $n$ which is the number of persons who will start the information propagation process.

As a result, there are three distinctive and mutually exclusive groups:

I. People who never heard the news (uninformed – "ignorant") $= X_t$, ($X_t = N$ for $t = 0$)

II. People who spreading the news ("spreaders") $= Y_t$, ($Y_t = n$ for $t = 0$)

III. People who heard the news and became inactive ("stifles") $= Z_t$, ($Z_t = 0$ for $t = 0$)

The total population $(N+n)$ at every instant of the process is:

$$X_t + Y_t + Z_t = N + n = S$$

There is a distinct difference between a typical epidemic model (see Bailey (1960) and Kentall (1956) ) and a model of diffusion that lies in the transition of a person from group (I) to group (II). A person from group (II) can "meet" two types of persons in his person-to-person contacts. It must be kept in mind that in this simplified version of the stochastic model, a fundamental assumption has been made that a person becomes a "stifler" if he encounters another person who knew the information. Thus, if a person from group (II) meets a person from the same group two units are lost from group (II). If he meets a person from group (III) then he stops transmitting the information and thus one unit is lost from group (II).

According to the above assumptions the rates of transfer are taken to be proportional to the meeting frequencies of different groups' members. Thus the first case is proportional to $0.5 \, Y_t \, (Y_t - 1)$ and the second case to $Y_t \, Z_t$. The possible transitions are given in Table 2 for the time $(t, t + dt)$.

**Table 2.** Transitions for person-to-person contact.

| TRANSITIONS | INFORMATION DISSEMINATION |
|---|---|
| $(X_t, Y_t, Z_t)$ fi $(X_t - 1, Y_t + 1, Z_t)$ | $X_t\ Y_t\ dt$ |
| $(X_t, Y_t, Z_t)$ fi $(X_t, Y_t - 2, Z_t + 2)$ | $\frac{1}{2} Y_t\ (Y_t - 1)\ dt$ |
| $(X_t, Y_t, Z_t)$ fi $(X_t, Y_t - 1, Z_t + 1)$ | $Y_t\ Z_t\ dt$ |

### Formulating the Propagation Assumptions as Markov Chain

According to the above assumptions, the system of information propagation to a Markov chain could then be transformed by defining the states as follows:

State 0:　when a handicapped from group $X$ meets a handicapped for group $Y$.
　　　　　Then　decreased　$X$ group and increased $Y$ (1 unit)
Stage 1:　when a handicapped from group Y meets a handicapped for group Y.
　　　　　Then decreased $Y$ group (2 units) and increased Z group (2 units).
Stage 2:　when a handicapped from group $Y$ meets a handicapped for group Z.
　　　　　Then decreased $Y$ group (1 unit) and increased Z group (1 unit).

The probability $p$ of a specific "knowledge-spreader" transmitting the information, given that he meets an "ignorant", has been assumed constant and equal to 1 in the preliminary exposition; now it is taken as ranging from 0 to 1. The probabilities for all states are a function of the density of the handicapped in the population, the type of news that they are spreading, their participation in organizations and their communicational habits. Thus, if two "knowledge-spreaders" meet, then the probability that one of them does not transmit the news is $(1 - p)$, both not transmitting is $(1 - p)^2$, and both transmitting to each other is $p\ (1 - p)$. The states and the transition probabilities are given in table 3.

**Table 3.** States and transition probabilities

| States | Transitions | Probabilities |
|---|---|---|
| 0 | $(X_t, Y_t, Z_t)$ fi $(X_t - 1, Y_t + 1, Z_t)$ | $p$ |
| 1 | $(X_t, Y_t, Z_t)$ fi $(X_t, Y_t - 2, Z_t + 2)$ | $(1-p)^2$ |
| 2 | $(X_t, Y_t, Z_t)$ fi $(X_t, Y_t - 1, Z_t + 1)$ | $p(1-p)$ |

Where: $X_t \neq 0, Y_t \neq 0, Z_t \neq 0$ and $X_t + Y_t + Z_t = N + n$

Based on Markov theory, information propagation states can be represented by a finite number of transition probabilities. As a result the information propagation has a finite number of possible states with a definite transition probabilities $p_{ij}$, which in state

$j$ will move to state $i$ after one unit of time (e.g. 1 week). A state vector can give the probability of each state $i$. and a transition matrix can be made of the state that reflects the transition probabilities $p_{ij}$. The state vector would present the probability of a given step being selected at the end of first time unit (1 week), at the end of second time (2 weeks) and so on.

## Numerical Application

The application that initiated the work presented in this paper, involves the estimation of the time-lag for a number of people knowing about the service before its actual operation. The purpose of this application focuses on two main issues; first to present the steps for applying the methodology framework in real word problems and second to present the formulation approach in order to be an easy to handle process for future users.

Assuming, that Athens International Airport has established a new ground transportation service directed to the elderly and handicapped and referred to provide transit services from/to airport. The potential message that the marketing department would like to transmit into the handicapped community includes: "The airport operator has established a new ground transportation service. If a handicapped want to transmit from/to airport could call to this ... phone number in order to fix the transportation arrangements".

According the above modeling framework, the first step is to create the transition matrix. When the information propagation procedure starts ($t = 0$) the total population size $N$ should be known, as well as the size of population elements that are spreading the news $(Y)$. In a questionnaire survey in which 100 handicapped people from the major relevant organizations in Athens that participated in this effort were interviewed and the population behavior and habits were obtained and recorded. This sample constitutes the *10%* of total population size (organization members) that had chosen to participate in this research. Thus, the first state the vector $(U)$ is presented in table 4 ($Y_{t=0}$ was *10%*).

**Table 4.** First state vector for the handicapped application

$$
U = \begin{array}{c|c}
\text{State} & \\
0 & 0.9 \\
1 & 0.1 \\
2 & 0.0 \\
\end{array}
$$

Analyzing the survey results, the probability of a specific "knowledge-spreader" transmitting the information was estimated *(p= 0.4)* and a transition matrix is constructed as follows:

**Table 5.** Transition matrix for handicapped group

$$
P = \begin{array}{c|ccc}
\text{State} & 0 & 1 & 2 \\
\hline
0 & 0.50 & 0.40 & 0.10 \\
1 & 0.40 & 0.36 & 0.24 \\
2 & 0.10 & 0.24 & 0.66 \\
\end{array}
\qquad \text{Where} \sum_{i,j}^{2} p_{ij} = 1
$$

The above matrix indicates that in the time unit $t=0$ a handicapped person from group $Y$ has a *36%* chance to meet a handicapped person for group $Z$ and *24%* to meet a handicapped person from group $Y$. Alternatively, could be depicted graphically the transition matrix, as below:

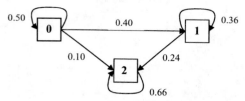

**Figure 1**. The state transition diagram

After one week (the first step of the Markov process) the new subgroups proportions are calculated and the next state vector is developed as follows:

**Table 6.** Second state vector for handicapped application

$$U^1 = \begin{array}{c} \text{State} \\ \begin{array}{c|c} 0 & 0.49 \\ 1 & 0.396 \\ 2 & 0.114 \end{array} \end{array}$$

The above matrix gives the proportion of each population subgroup after *1* week. Therefore, after one week only 49% will know the new information and approximately 40% will be the "knowledge-spreaders", who transmit the news to the others. Based on the Markov process assumptions, the proportions of population group during 2, 3, 4 and 12 weeks is calculated as presented in the follow matrices.

**Table 7.** States vector for handicapped application

$$U^2 = \begin{array}{c} \text{State} \\ \begin{array}{c|c} 0 & 0.415 \\ 1 & 0.366 \\ 2 & 0.219 \end{array} \end{array} \quad U^3 = \begin{array}{c|c} 0.376 \\ 0.350 \\ 0.274 \end{array} \quad U^4 = \begin{array}{c|c} 0.355 \\ 0.342 \\ 0.302 \end{array} \quad \cdots \quad U^{12} = \begin{array}{c|c} 0.333 \\ 0.333 \\ 0.333 \end{array}$$

As a result, the application of the Markov process indicates that in 12 weeks all the rows of the matrix have identical entries, which means that all proportions of population groups are equal, and as depicts in Figure 2. In other words, the 66% of the total population studied would have been informed about the new airport service. Moreover, the Markov process in the 12$^{th}$ state gives a limiting transition probability (steady-state probability) that is independent of the initial state.

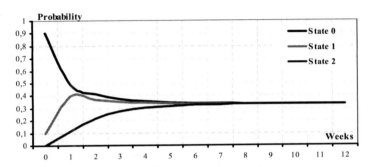

**Figure 2.** Probabilities for the three states transitions during 12 weeks

## Conclusions

The process of information dissemination, which is discussed in this paper, deals with the person-to-person contact within a special population group that in many cases constitutes the optimum promotion strategy. In this manner, the news of starting a new ground transportation service at airports will reach the entire target population through the dynamics of information diffusion. Given the complexity of real-world problems, the information diffusion is dependent on the probability of transmitting news, the communication habits of under study population and the time it takes for the news to propagate throughout the community.

This paper presents a methodology and framework to estimate the time required for $N$ numbers of members of the population subgroups to be informed about a new service, before the service starts its operation. The proposed stochastic formulation approach is focused on handicapped community and three separate subgroups are distinguished in order for the required time of service commencement to be announced and all members of the population informed, before the service actually starts its operation. Key elements for using this formulation are: (a) to formulate the propagation assumptions as Markov chain according to propagation structure for understudy population; (b) to acquire the first step transition probabilities that constitute the input for the next states; and (c) to calculate the transition probabilities over the time.

It has been shown that the proposed formulation could provide a manageable and efficient tool for the users to adopting both low cost marketing strategies, and effective promotion policies; taking into account the characteristics of closed population systems. The goals of this research are to use the easy-to-handle Markov formulation that adopts three group units with different transmitting characteristics to provide real-life application of this methodology and framework for focused special airport ground transportation services.

## References

Anselmi K. (2000). *A Brand's Advertising and Promotion Allocation Strategy - The Role of the Manufacturer's Relationship with Distributors as Moderated by Relative*

*Market Share*, Journal of Business Research, Volume 48, Issue 2, pp 113-122

Bailey, N.T.J. (1960). *"The mathematical theory of epidemics"*, Chapter 6, Griffin, London.

Bauman A.E., Bellew B., Owen N. and Vita P. (2001). *Impact of an Australian mass media campaign targeting physical activity in 1998*, Volume 21, Issue 1, pp. 41-47

Bhat, U.N., and Miller G.K. (2002). *Elements of Applied Stochastic Processes*, 3rd ed., Wiley, NY.

Bronnenberg, B.J. (1998). *Advertising frequency decisions in a discrete Markov process under a budget constraint*, Journal of Marketing Research 35 (3), 399–406.

Cheng, F. and Sethi, S.P. (1999). *A periodic review inventory model with demand influenced by promotion decisions*, Management Science 45 (11), pp. 1510–1523.

Daley, D.J., and Kendall, D.G. (1956). "Stochastic Rumors", J. Institute of Mathematic Applications, 42-55.

EUROSTAT (2001). "European Community Household Panel", wave 8.

Hillier, F., and Lieberman, G. (2005). "Chapter 16: Markov Chain", *Operations Research*, Mc Graw-Hill, International Edition, 16.1-16.4.

Kentall, D.G. (1956). *Deterministic and Stochastic Epidemics in closed Populations*, Proceedings of 3rd Symposium on Mathematical Statistics and Probability, Berkeley, 4.

Kurataa, H., and Liub, J. J. (2006). *Optimal promotion planning-depth and frequency-for a two-stage supply chain under Markov switching demand*, European Journal of Operational Research Volume 177, Production Manufacturing and Logistics, Issue 2 , 1026-1043.

McClure, J., and Kumcu, E. (2006). *Promotions and product pricing: Parsimony versus Veblenesque demand*, Journal of Economic Behavior & Organization, article in Press.

MacKay, K. J., and Smith, M. C. (2006). *Destination advertising age and format effects on memory*, Annals of Tourism Research, Volume 33, Issue 1 , pp 7-24

Mela, C. F., Gupta, S. and Lehmann, D.R. (1997). *The Long-Term Impact of Promotion and Advertising on Consumer Brand Choice*, Journal of Marketing Research 34, 248–261.

Jones, J. P. (2004). *Advertising and Promotion*, Encyclopedia of International Media and Communications, pp. 7-15

Palmer, K., Dessouky, M., and Abdelmaguid, T. (2004). *Impacts of management practices and advanced technologies on demand responsive transit systems*, Transportation Research Part A: Policy and Practice, Volume 38, Issue 7, 495-509.

Pascual, M., and Cantarerol, D., (2006). *Socio-demographic determinants of disabled people: An empirical approach based on the European Community Household Panel*, Journal of Socio-Economics Volume 36, Issue 2, 275-287.

Rekieka, B., Delchambrea, A., and Aziz Salehb, H. (2006). Handicapped *Person Transportation: An application of the Grouping Genetic Algorithm*, Engineering Applications of Artificial Intelligence, Volume 19, Issue 5, 511-520.

Sethuraman, R., and Tellis, G. J. (1991). *An Analysis of the Tradeoff Between*

*Advertising and Price Discounting*, Journal of Marketing Research, 160–174.

Smith, M. C., and MacKay, K. J. (2001). *The Organization of Information in Memory for Pictures of Tourist Destinations: Are There Age-Related Differences?*, Journal of Travel Research 39, 261–266.

Sogomonian, A.G. and Tang, C.S., (1993). *A modeling framework for coordinating promotion and production decisions within a firm*, Management Science 39 (2), 191–203.

Stephanis, B.K, and Dimitriou, D.J. (2006). *Information propagation: informing special population subgroups about new transportation services*, Proceedings of International Conference on Transportation Research, Hellenic Institute of Transportation Engineers, CD proceedings.

Fam, K.-S., and Yang, Z. (2005). *Primary influences of environmental uncertainty on promotions budget allocation and performance: A cross-country study of retail advertisers*, Journal of Business Research, Volume 59, Issue 2, pp 259-267

# Simulation of Quadruple IFR Arrival Runways and End-Around Taxiways at George Bush Intercontinental Airport

M. T. McNerney, PhD, PE (Member)[1], and B. Hargrove[2]

[1]DMJM Aviation, 1200 Summit Ave, Suite 320, Fort Worth, TX 76102; PH (817) 698-6830; FAX (817) 698-6802; email: mike.mcnerney@aecom.com
[2]TransSolutions Inc, 1400 Trinity Blvd, Suite 200, Fort Worth, TX 76155; PH (817) 359-2958; FAX (817) 359-2959; email: bhargrove@transsolutions.com

*Abstract*

At the beginning of its Master Plan, Houston George Bush Intercontinental Airport had two pressing questions, did the airport need four simultaneous instrument flight rules (IFR) landing runways and would end-around taxiways be a cost effective improvement? During the alternatives analysis it was determined that quadruple IFR landing runways were not only part of the preferred alternative, but aircraft simulation analysis also proved that triple IFR landing runways would not reach the desired hourly capacity goals or annual delay reduction goals. Simulation analysis of the implementation plan showed that a fourth IFR arrival runway should be the first of two needed runways and that an end-around taxiway was needed even before the next runway can be constructed.

*Background*

Based upon Airport Council International 2005 reported data, Houston George Bush Intercontinental Airport is currently the sixth busiest airport in the world in terms of aircraft operations (ACI-NA 2005). In November 2004, George Bush Airport became only the third US airport to have three simultaneous IFR landing runways. Currently, no airport has four simultaneous and independent IFR landing runways. The FAA has given approval for the concept of quadruple IFR runways and the future runways are in the master plans for Denver International and Dallas/Fort Worth International airports.

The Houston Airport System is the owner/operator for the City of Houston of three airports: William P Hobby Airport (HOU); Ellington Field (EFD); and George Bush Intercontinental Airport (IAH). In May 2004, the Houston Airport System (HAS) contracted with DMJM Aviation to conduct a Master Plan for the George Bush Intercontinental Airport. George Bush Intercontinental Airport is the primary hub of Continental Airlines and has been growing steadily except for the year 2001 decline. Two of the primary questions in the Master Plan were to determine if quadruple IFR landing runway capacity was needed in the 20-year planning time frame and if "end-around taxiways" (also called "perimeter taxiways") would be a cost effective improvement.

This paper presents the results of extensive simulation during the alternative analysis and implementation planning phases of the master plan. The results show that Houston George Bush does indeed need a fourth simultaneous IFR landing runway and that end-around taxiways should be built as soon as possible.

### Forecast of Operations

At the start of the Master Plan in May of 2004, the airport's fifth runway was under construction, which would provide three parallel and independent runways that could be used for simultaneous IFR landings in very poor weather. With IAH expected to achieve triple IFR landings it was necessary to use this 5-runway capacity as a baseline for future analysis. A 20-year forecast had been completed based upon 2003 actual operations, which has a fleet mix of 30% operations of regional jet aircraft growing to 40% at the end of 20 year planning period. However, a check of 2004 actual data and interviews with Continental Airlines and Continental Express, the regional jet affiliate owned by Continental Airlines, indicated that 40% regional jet operations would be reached at the airport in 2005.

With the same number of total passengers at the airport and a fleet mix change that increases the percentage of 50-70 seats regional jets would mean a higher number of aircraft operations. An extensive aircraft operations forecast was completed using a new fleet mix in cooperation with and reviewed by Continental Airlines. Figure 1 shows the 20-year forecast of annual operations developed for the master plan which tops out at 880,000 annual operations at the year 2025. This new forecast was a 10% percent increase for year 2010 over the previous 2003 forecast and was within 5% of the 2003 FAA Terminal Area Forecast for IAH.

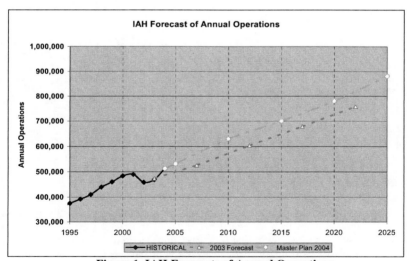

**Figure 1. IAH Forecasts of Annual Operations**

After the baseline forecast was validated by the Airport and reviewed by Continental Airlines, the annual forecast was then turned into a flight schedule that represented the average day of the peak month. The peak month for past years has been July and the average weekday in July 2004 was grown based upon actual fleet forecast from Continental Airlines and historic data from other tenant airlines. Based upon reviews of several consultants, the FAA, and Continental Airlines, the Master Plan team was confident that the hourly schedules for 2005, 2010, 2015 and 2025 were the reasonable daily schedules for IAH. Figure 2 shows the rolling 60 minute schedule of operations for each of the years. An interesting point in this analysis was that not only was there going to be a significant growth in peak hour operations, but the 2005 peak hour was roughly equivalent to the 2025 slack hour or trough point in the graph of operations. Although, it is not cost effective to design for the absolute peak hour demand, it was a goal to achieve an hourly capacity of 210 combined takeoff and landing operations per hour.

The issue of "de-peaking" the flight schedule to reduce congestion was discussed with Continental Airlines. The 2025 schedule is not a true depeaked schedule but represents a realistic schedule without large troughs or quiet periods in the schedule. Much of the schedule growth occurs in the trough periods and results in a schedule that is partially depeaked, but realistic.

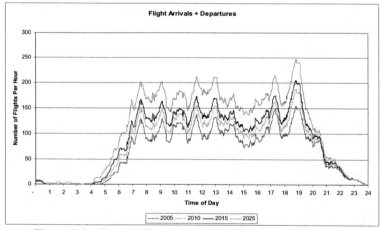

**Figure 2. Rolling 60 Minute Forecast of Scheduled Operations**

*Airside Alternatives Analysis*

George Bush Intercontinental Airport was first opened with two runways in an "Open V" configuration then expanded to a total of three east-west runways used mostly for landing and two northwest southeast runways used primarily for takeoffs. The capacity of the existing airfield cannot accommodate the expected demand in 2015, much less 2025. Thus, thirteen airfield alternatives were developed and analyzed. The alternatives consisted of three different configurations with cases of triple or quadruple IFR landing runways for each configuration. Most alternatives were in the open V configuration, with alternatives of all parallel east west runways configurations and all northeast southwest configurations. First, each of the 13 alternatives was analyzed for hourly IFR and VFR capacity (with high-level analysis using the FAA's Airfield Capacity Model), environmental impact, construction cost, and other criteria. Five Alternatives shown in Figure 3 were selected for further analysis using detailed simulation analysis to calculate delay. Although none of the triple IFR alternatives met the hourly capacity goals, the best triple IFR option, Alternative 1.2, was carried into the detailed analysis to determine if delay goals could be reached with triple IFR runways.

The FAA's Airport and Airspace Simulation Model, SIMMOD, was used for the detailed simulation analyses. Developed by the FAA over 20+ years, SIMMOD has been used worldwide for a variety of airfield and airspace capacity/delay studies. The model is extremely flexible, allowing analysts to represent very broad airspace areas down to individual aircraft gates and parking constraints. Output is provided for each aircraft such that hourly throughput, travel times, and delay times are easily obtained. The IAH SIMMOD model included very detailed airfield with only final approach and initial departure headings in the airspace. This approach was suggested by the FAA so that the Airport Master Plan would focus on airfield developments over which the Airport has authority while the FAA will then conduct an airspace redesign effort to effectively move aircraft traffic to/from the Airport's preferred runway layout.

Goals were established for annual average arrival delay, annual average departure delay, and annual average delay per aircraft operation. Being a hub airport, the goals for annual average delays was set for 3 minutes for arrival, 6 minutes for departure, and 4 minutes per average operation.

The airside simulation was conducted by TransSolutions under the direction of DMJM Aviation for the five alternatives. Simulation runs were performed in east- and west-flow and under IFR and VFR traffic conditions. Results reported are the average of ten simulations as is standard procedure in SIMMOD analyses. Detailed runway procedures for each wind/weather scenario of every alternative were developed and reviewed with the local FAA IAH air traffic controllers. The interaction of Runways 15-33 departures with the east-west arrival runways was particularly critical in the analysis because the Runway 15L and 15R departure release is dependent upon protecting for the missed approach off Runway 27 and potential Runways 27C and 27L. Discussions with local FAA controllers established the runway procedures for the proposed runway locations to ensure their accuracy and reasonableness.

Measurements of the simulations were taken of the operations per each hour, the average delay by hour, the peak hourly capacity, the peak hourly delay, the average taxi time, the average and maximum aircraft in the departure queue and several other metrics. Using the four wind/weather configurations simulated, a weighted average was calculated to derive average annualized delay.

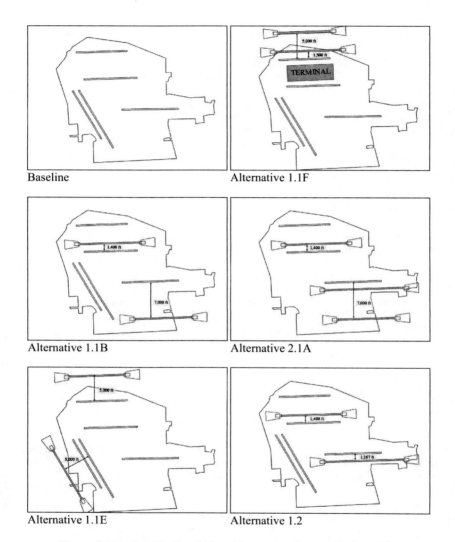

Baseline

Alternative 1.1F

Alternative 1.1B

Alternative 2.1A

Alternative 1.1E

Alternative 1.2

**Figure 3. Baseline Case and Five Alternatives Selected for Simulation**

Because visual approaches for three arrival runways could only be flown when the ceiling is at least 10,000 feet and the visibility at least 8 miles, there is a considerable amount of time the airport is using instrument flight rules. Although part of the time under IFR rules, visibility is such that there is a reduced interaction from landing aircraft requiring protected space from departing aircraft. While this would reduce some of the calculated departure delay, the arrival spacing is not significantly different such that the calculated arrival delays are reasonable.

Since the alternative layouts are being considered for the long-term (20-year) development plan, the alternative analysis tested each alternative's ability to accommodate the 2025 forecast demand. Using the average day peak month schedule, the alternatives were all tested against the goals for hourly capacity and average delay.

The results of the simulations of the alternatives are shown in Table 1. The Baseline case has annual arrival delays of 15.70 minutes and annual departure delays of 29.10 minutes and an average annual delay per aircraft of 22.40 minutes. These are unacceptable delays and for comparison purposes the most congested airports in New York and Philadelphia only have average annual delays per aircraft of 10 minutes (NPIAS 2005). If you notice the comparison of IFR and VFR delays, the base case has average delays of over 50 minutes for IFR conditions in east-flow and in west-flow it has over 30 minutes for arrival and 60 minutes for departure. These IFR delay conditions would result in a total gridlock at the airport based upon runway capacity, assuming the airspace could accommodate the traffic to/from the airfield.

Alternative 1.2 was judged as the best of the alternatives that has only a triple simultaneous IFR landing capacity. This alternative was included into the simulations after the first screening to determine if a triple IFR solution could meet the hourly capacity and annual delay goals. The simulations results show that the annual arrival delay is 8.05 minutes compared to a goal of 3 minutes for a hub airport. The alternative did achieve the departure delay of 6 minutes with a 4.24 average departure delay. The alternative could not achieve a 3.0 minute arrival delay even in VFR conditions and has average arrival delays of 13 and 14 minutes in west and east flow in IFR conditions. From the simulation results, all of the quadruple IFR landing alternatives are clearly superior to the triple IFR alternative and the triple IFR alternative can not handle the 880,000 annual traffic demands with the acceptable IFR arrival delays.

The remaining four alternatives met the goals for annual arrival delay, annual departure delay and average delay per aircraft. Alternative 2.1A is an all east-west runway configuration requiring the building of 3 new runways and the closing of two existing runways. It has the best arrival delay at 2.01 minutes of delay. However, it also has the worst departure delay of the four quadruple IFR alternatives at 5.51 minutes.

Alternative 1.1E has one new 15-33 departure runway and a parallel 9-27 arrival runway. This alternative expands the "open V" configuration of the airport outward and has the least runway interaction. This alternative has the lowest departure delay and the lowest average delay per aircraft. With respect to simulation results, this alternative is the clear winner. However, all four of the alternatives reached our goals for delay and the results are fairly close with results of 3.19, 3.56, 3.76, and 3.99 for average delay per aircraft.

**Table 1. Simulation Results of Alternative Analysis (in minutes)**

| | NUMBER OF ARRIVAL RUNWAYS | FLOW | WEATHER | SIMMOD OPS | AVERAGE ARRIVAL DELAY | AVERAGE DEPART DELAY | ANNUAL ARRIVAL DELAY | ANNUAL DEPART DELAY | ANNUAL DELAY PER AIRCRAFT |
|---|---|---|---|---|---|---|---|---|---|
| 1.1.B | 4 | EAST | VFR | 204 | 3.2 | 2.5 | 3.03 | 4.96 | 3.99 |
| | | | IFR | 189 | 6.2 | 5.9 | | | |
| | | WEST | VFR | 219 | 2.2 | 4.5 | | | |
| | | | IFR | 196 | 3.0 | 6.6 | | | |
| 1.2 | 3 | EAST | VFR | 204 | 7.1 | 1.5 | 8.05 | 4.24 | 6.15 |
| | | | IFR | 185 | 14.6 | 4.8 | | | |
| | | WEST | VFR | 201 | 3.6 | 3.0 | | | |
| | | | IFR | 183 | 13.3 | 7.5 | | | |
| 2.1.A | 4 | EAST | VFR | 197 | 1.3 | 5.2 | 2.01 | 5.51 | 3.76 |
| | | | IFR | 195 | 3.1 | 6.0 | | | |
| | | WEST | VFR | 197 | 1.3 | 5.2 | | | |
| | | | IFR | 195 | 3.1 | 6.0 | | | |
| 1.1.E | 4 | EAST | VFR | 219 | 2.8 | 2.3 | 2.43 | 3.95 | 3.19 |
| | | | IFR | 198 | 3.5 | 6.4 | | | |
| | | WEST | VFR | 227 | 1.6 | 2.1 | | | |
| | | | IFR | 195 | 3.2 | 6.9 | | | |
| 1.1.F | 4 | EAST | VFR | 204 | 1.2 | 4.1 | 2.16 | 4.95 | 3.56 |
| | | | IFR | 190 | 3.1 | 6.8 | | | |
| | | WEST | VFR | 208 | 1.6 | 3.5 | | | |
| | | | IFR | 188 | 3.2 | 7.2 | | | |
| BASE LINE | 3 | EAST | VFR | 187 | 4.7 | 6.1 | 15.70 | 29.10 | 22.40 |
| | | | IFR | 167 | 50.2 | 54.4 | | | |
| | | WEST | VFR | 193 | 4.1 | 12.7 | | | |
| | | | IFR | 171 | 31.7 | 67.4 | | | |
| | | | | | | | GOAL 3.00 | GOAL 6.00 | GOAL 4.00 |

The selection of the preferred alternative was the result of a full evaluation of the four acceptable alternatives based upon simulated capacity/delay results, taxi time and distance, cost of construction, land acquisition, community impact, noise and other environmental factors. These four alternatives were studied for each of evaluation criteria and the results were briefed to the public, the stakeholders of the airport, the airport staff, and the FAA.

The comparison of the four acceptable alternatives is shown in Table 2. Each alternative had unique factors to evaluate. Alternative 2.1A would close two existing departure runways. Alternative 1.1F would work better if a new terminal were location to the north of the existing terminals. It would also be the option of choice if the airport were considering the need for a new architectural statement terminal building such as Denver International's Jeppesen Terminal building. Alternative 1.1E would require expanding the airport to the West. Alternative 1.1B just barely meet all goals but had the least land acquisition cost and community impact.

**Table 2. Evaluation Factors of Alternatives**

|                      | 1.1B                         | 1.1E                         | 1.1F                         | 2.1A                         |
|----------------------|------------------------------|------------------------------|------------------------------|------------------------------|
|                      | South Quad                   | Super Vee                    | North Quad North Terminal    | All East West                |
| Runway Capacity      | Good 212 Ops/hr              | Best 240 Ops/hr              | Good 218 Ops/hr              | Good 218 Ops/hr              |
| Annual Delay         | Good 3.99 minutes            | Best 3.19 minutes            | Good 3.56 minutes            | Good 3.76 minutes            |
| Construction Cost     | Best $461 Million            | Good $499 Million            | Good $518 Million            | Worst $707 Million           |
| Land Acquisition     | Best $53 Million             | Worst $124 Million           | Poor $102 Million            | Good $55 Million             |
| Community Impact     | Best 1,932 pop within 65 Ldn | Poor 2,497 pop within 65 Ldn | Worst 5,170 pop within 65 Ldn | Poor 2,212 pop within 65 Ldn |

After integration of the airside analysis with the terminal analysis and ground access analysis the airport selected alternative 1.1B as the preferred alternative. As shown in Figure 4, Alternative 1.1B has a new Runway 9R-27L located 7000 feet (2133 m) south of existing Runway 9-27. This runway would be used primarily for a quadruple IFR landing runway and must be spaced at least 5000 feet (1524 m) from Runway 9-27 for independent operations. In addition the alternative includes a new Runway 8C-26C, 12,000 feet (3657 m) in length, located 1417 feet (432 m) north of existing Runway 8R-26L. This runway would be used for departures only and would be dependent with Runway 8R-26L during IFR operations but would be independent during VFR operations.

Another feature of Alternative 1.1B is the end-around taxiways around Runway 8R-26L. The primary feature of these is not to have to cross Runway 8R-26L. The secondary purpose is that they provide additional queuing space for departing aircraft.

After the decision was made selecting Alternative 1.1B as the preferred alternative, approximately 50 additional simulations were completed to analyze the implementation of the alternative and choose which runway should be built first and when the planning and construction should begin.

*Airside Implementation Analysis*

Simulation of the Baseline condition and the final build out of two new runways had already been accomplished for the year 2025 at 880,000 annual operations. However, in order to build a demand delay curve and determine which runway and end-around taxiways should be built first each increment was simulated in a 2010, 2015, 2020 and 2025 year as necessary. Each case had to be simulated in IFR and VFR, and east and west flow. Many cases had to be simulated several times in order to optimize runway assignments. The percentage of takeoffs on any particular runway was a manual preset on each run of the simulation. Often a small tweak of the percentage was necessary to balance the runways of the airfield layout for minimum delay.

**Figure 4. The Preferred Alternative Number 1.1B**

Because it is not possible to perform an environmental impact statement (EIS) and design and build a new runway all in 5 years, the first enhancements included a new cross taxiway between the north and south aprons which had already been planned to relieve bidirectional traffic congestion on that taxiway. Two runway crossing taxiways were added to allow aircraft to cross Runway 15L to queue for departure for Runway 15R between the two runways. Also simulated was the end-around taxiways designed to allow aircraft to taxi around existing Runway 8R-26L. These end-around taxiways would be the primary departure routes for future Runway 8C-26C and the primary taxi-in route for existing Runway 8L-26R. Because of the departure queuing deficiency of the existing Baseline case it was decided to keep the end-around taxiways in place even before future Runway 8C-26C was constructed. From the results this turned out to be fortuitous decision.

To annualize delay, one must take into account the annual percentage of time in IFR versus VFR operations and in east- and west-flow. Currently, west-flow capacity is higher, so the FAA tower controllers prefer to operate the in west flow, even if a 5-10 knot tailwind exists. Wind analysis shows that an east-flow is the more favorable wind condition throughout the year. With the new layout, the highest capacity configuration may be different, so the annualized delays were calculated for operating in east during calm winds (up to 6 knots of tailwind) and operating in west during calm winds. Therefore, in Table 3 results are calculated for maximum east and maximum west for comparison purposes. In the overall analysis, the average of the two conditions is a good approximation.

**Table 3. Annualized Delay in Minutes for Implementation Planning**

| | 2010 | 2015 | 2020 | 2025 | 2010 | 2015 | 2020 | 2025 |
|---|---|---|---|---|---|---|---|---|
| **Average Annualized Delay Per Flight** | Max East Configuration | | | | Max West Configuration | | | |
| Baseline | 4.16 | 6.95 | 11.80 | 28.03 | 4.03 | 6.52 | 12.46 | 26.41 |
| End-around Taxiways | 3.37 | 4.23 | 6.14 | 21.63 | 3.32 | 4.28 | 5.97 | 20.89 |
| Quad Arrival RWY | 2.23 | 3.02 | 4.60 | 8.70 | 2.18 | 3.00 | 4.53 | 10.13 |
| Departure RWY | 1.89 | 2.62 | 4.74 | 8.54 | 1.89 | 2.61 | 4.71 | 9.94 |
| Full Alternative 1.1B | | | 2.56 | 4.12 | | | 2.95 | 4.07 |
| **Average Annualized Arrival Delay** | Max East Configuration | | | | Max West Configuration | | | |
| Baseline | 2.54 | 3.14 | 7.00 | 21.49 | 2.78 | 3.47 | 6.76 | 18.72 |
| End-around Taxiways | 2.43 | 2.98 | 3.91 | 6.81 | 2.46 | 2.98 | 3.97 | 7.06 |
| Quad Arrival RWY | 1.40 | 1.70 | 1.97 | 2.56 | 1.22 | 1.46 | 1.80 | 2.35 |
| Departure RWY | 2.42 | 3.15 | 3.99 | 7.71 | 2.29 | 2.92 | 3.80 | 7.22 |
| Full Alternative 1.1B | | | 1.7 | 4.08 | | | 1.7 | 3.36 |
| **Average Annualized Departure Delay** | Max East Configuration | | | | Max West Configuration | | | |
| Baseline | 5.77 | 10.76 | 16.60 | 34.56 | 5.28 | 9.56 | 18.15 | 34.11 |
| End-around Taxiways | 4.31 | 5.48 | 8.37 | 36.45 | 4.18 | 5.57 | 7.98 | 34.71 |
| Quad Arrival RWY | 3.06 | 4.35 | 7.24 | 14.84 | 3.15 | 4.55 | 7.26 | 17.91 |
| Departure RWY | 1.37 | 2.09 | 5.50 | 9.36 | 1.48 | 2.31 | 5.62 | 12.67 |
| Full Alternative 1.1B | | | 3.42 | 4.15 | | | 4.20 | 4.78 |

The most surprising result was in comparing the Baseline results to the enhanced Baseline by adding only a few taxiways and end-around taxiways but no new runway. In the Baseline condition, the FAA tower does not like to use the new north Runway 8L-26R for departures because of neighborhood noise concerns and because of the belief that dedicating the runway to arrivals only would provide the most capacity. However, in the analysis of the condition with end-around taxiways, by adding a few departures on the north runway and using the end-around taxiways as the queuing path for aircraft departing on Runway 8R-26L, there was a significant reduction in departure and overall delay with a small reduction in arrival delay. The reduction in delay was much more significant in 2015 and 2020 but did not hold true for 2025.

The difference in departure delay in the year 2020 (780,000 annual operations) from 17 to 8 minutes was so dramatic it was hard to accept as factual. Not until the animations of the simulations were viewed and the difference in taxi-out patterns was observed could it be understood. Without the end-around taxiways, the taxiways adjacent to the terminal are used as queuing space and aircraft arriving to the terminal actually have to cross Runway 8R-26L at midfield to minimize the congestion. With end-around taxiways, the aircraft queue on the end-around taxiways and the arriving aircraft taxi in around Runway 8R-26L and the terminal area taxiways are clear of congestion.

Because the end-around taxiways could be built sooner than a new runway and the delay reduction was so significant even without a new runway, they were considered as the first phase of new construction in the implementation plan.

The second step was to determine if the first new runway was to be a quadruple arrival runway or a new departure only runway. The pervasive thinking was that the triple arrivals would provide good service for awhile but the departure delays were starting to be a big problem. If no new construction were undertaken, the 2010 schedule (700,000 annual operations) would result in 10 minutes of annual departure delay per aircraft. Therefore, the generally accepted premise was that a departure runway would be the first runway needed.

However, the results of the simulation surprised most everyone. Figure 5 shows that if the end-around taxiways are built first, there is a significant reduction in departure delay. The Baseline departure delay exceeds 6 minutes at 630,000 annual operations, but if the end-around taxiways are built, the departure delay will not exceed 6 minutes until about 720,000 annual operations. Building both end-around taxiways and the arrival runway extends the departure capacity to 750,000 annual operations, while building the end-around taxiways and the departure runway extends the departure capacity only to 780,000 annual operations. Based only on departure delay, both new runways are needed by 780,000 annual operations.

Building the departure runway first instead of the arrival runway first only provides a reduction below the departure delay goal of 6 minutes per aircraft only between 750,000 and 780,000 operations. However, building the arrival runway first provides a significant reduction in average arrival delay especially in IFR operations.

As shown in Figure 6, the end-around taxiways and the departure runway do not provide any significant reduction in arrival delay. Arrival delay is measured as aircraft waiting to land and waiting for runway capacity and spacing while in the air. For consistency in the analyses, airfield taxi delays were not included in the arrival delay output analyses. The goal of 3 minutes average arrival delay is exceeded in the baseline condition at 680,000 annual operations. The only way to reduce this delay is to build the arrival runway at around 700,000 annual operations. The arrival delay is never exceeded again. The departure runway is then built to reduce the departure delay around 780,000 annual operations.

### Simulation Animation

The animation of the simulation runs proved to be a very valuable tool for analyzing and showing the results of the simulations. Figure 7 is screen capture of an animation playback of the Base Case scenario in West Flow in IFR conditions, at 780,000 annual operations level of demand at a busy time of the time of the day around 1930 hours. In Figure 7, the red aircraft symbols indicate aircraft stopped that are mostly waiting in a departure queue or waiting to taxi to the departure queue showing heavy congestion in the taxiways near the terminal gates. In Figure 7, Runway 26R at the top of the figure is used exclusively for arrivals. The magenta aircraft symbols represent aircraft that are taxiing without restriction either for departure or arriving to a gate.

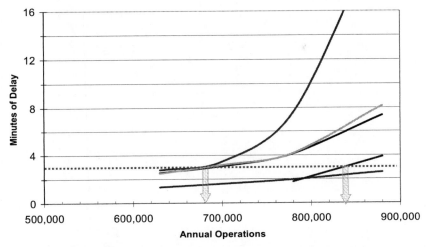

**Figure 6. Graph of Average Annual Arrival Delay Per Aircraft**

Figure 8 is a screen capture of the same time of day and same simulation conditions except that Runway 26R is used for both arrivals and some departures and the end-around taxiways are used for the departure path and queuing space for Runway 26L.

Comparing Figure 8 to Figure 7 there are significantly less red aircraft symbols representing less aircraft holding their taxiing for departure. There is less congestion around the terminal gates and more magenta aircraft symbols which indicate more aircraft taxiing in and out. The differences are even greater than what is shown in the screen capture because in each of the figures a single symbol can represent more than one aircraft at that node or point in the simulation and a number is placed beside the symbol. Comparing Figure 8 to Figure 7 there are no red aircraft symbols on the two parallel taxiways just north of the terminal gates in Figure 8 but there red aircraft symbols representing about 80 aircraft on the just those two taxiways in Figure 7. These figures are only a momentary snapshot in time, but the data and animation but show significant reduction in departure delay from having the end-around taxiway used as a departure queue.

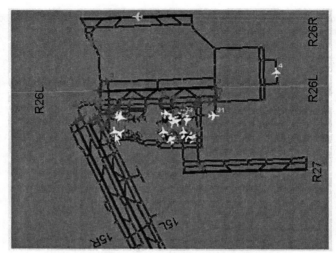

**Figure 7.  SIMMOD Visualization of Base Case**

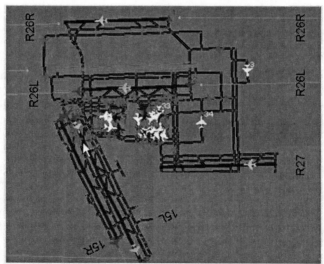

**Figure 8. SIMMOD Visualization of End-Around Taxiway used for Departure Queuing**

## Conclusions

The results of the SIMMOD simulations showed that triple IFR capacity even with two new runways would exceed the goal of an average of 3 minutes annual arrival delay and 4 minutes of average annual delay per aircraft before 880,000 annual operations at George Bush Intercontinental Airport. Therefore, four simultaneous IFR landing runways are needed at the airport. The results of the simulation also showed that the fourth arrival runway should be constructed as soon as 700,000 annual aircraft operations and should be the first of the two proposed new runways. The second new runway is needed primarily for departure capacity and should be construction by 780,000 annual operations.

The results of the simulations also showed quite emphatically, that an end-around taxiway around both ends of Runway 8R-26L should be constructed as a departure path as soon as 630,000 annual operations.

The FAA 2006 Terminal Area Forecast for IAH showed a more rapid increase in forecast operations such that the airport will reach 1.1 million annual operations by the year 2025 rather than the 880,000 annual operations used in the Master Plan analysis. This new FAA forecast is back up by rapid growth by Continental Airlines at IAH that has resulted in 41,000 new operations to the airport in each of the last two years.

The impact of this new forecast is that the master plan needs to be accelerated because the 700,000 annual operations trigger for completing the first runway could be reached in the year 2011 and the second new runway would be needed at 780,000 annual operations by the year 2015. With the average time required to conduct an environmental impact statement and then design and construct one new runway well over five years, there is compelling need to begin the process now.

## Acknowledgements

The authors wish to thank the Houston Airport System staff for their cooperation and support during the Airport Master Plan.

## References

ACI-NA 2005, 2005 North American Final Traffic Report: Total Movements, Airport Council International – North America, www.aci-na.org/asp/traffic.asp?art=217 Accessed July 24, 2006.

NPIAS 2005, National Plan of Integrated Airport Systems (NPIAS) (2005-2009) Report of the Secretary of Transportation to the US Congress, Table 4, (2002 Data).

# The Next Revolution in Air Transport - RNP

Kenneth A. Kvalheim, M. ASCE [1], and Hal Andersen, ATP [2]

[1] Naverus, Inc., Chief Designer, 20415 72nd Ave. S., Suite 300, Kent, WA 98032; PH (253) 867-3858; email: kkvalheim@naverus.com
[2] Naverus, Inc., Vice President Flight Operations, 20415 72nd Ave. S., Suite 300, Kent, WA 98032; PH (253) 867-3861; email: handersen@naverus.com

## Abstract

With airports and airways already experiencing congested conditions, and the demand for passenger and cargo transport expected to grow for the foreseeable future, a number of strategies have been developed to plan for increased safety, access, efficiency, and capacity at our nation's airports and around the world. One such strategy is the use of RNP (Required Navigation Performance) which is supported by the FAA and its *Roadmap for Performance-Based Navigation*. With the ability to follow a GPS-based route and use narrow, constant-width RNP containments, approaches and departures can follow precise, curvilinear paths to avoid difficult terrain and complex airspace. Airspace efficiency is improved, safety is enhanced, and airport capacity is increased by the implementation of procedures that use the technology already onboard most new aircraft. This paper details the benefits of RNP, describes the current process of change, provides a case study of RNP benefits, and looks into the future.

## Introduction

The manner in which aircraft use the global air space system is currently undergoing a major transformation that is considered by some to be a new revolution in aviation. Revolutionary changes occur when new technology matures to the point that it can provide simultaneous advancements in system efficiency, reliability, and safety. The "First Revolution" of flight came about as advancements in metallurgy and construction methods allowed aircraft builders to transition from wood and fabric biplanes to all-metal monoplanes in the late 1920's and early 1930's. Current aircraft construction methods and configurations are still largely based upon the changes that occurred as a result of this first revolution. The development of the jet engine represents what can be considered the "Second Revolution" of flight.

The "Third Revolution", which began in the mid-1980's, is the result of advancements in the computational power of aircraft automation and navigation systems and the introduction of satellite-based navigation systems (such as GPS and Europe's Galileo). These systems allow aircraft to automatically navigate through the airspace system with an unprecedented level of accuracy and precision. These advancements are specifically enabled through the use of Required Navigation Performance (RNP) systems and procedures.

RNP is a specialized form of RNAV (Area Navigation). An airplane, properly equipped for RNAV, can follow specifically planned routes using GPS and established waypoints for navigation, instead of ground-based radio transmitters. But with RNAV alone, the integrity and accuracy of the airplane's actual location is not reported. This is where RNP systems step in. With RNP, the onboard navigation systems provide real-time *monitoring and alerting* regarding the airplane's actual location. With RNP, the Required Navigation Performance is continually compared with the Actual Navigation Performance (ANP), so that there is assurance that the airplane is within the prearranged "containment" (or "tunnel" in the sky). If the ANP exceeds the RNP, the flight crew is immediately alerted, and the aircraft can climb or go-around, following a pre-programmed, safe extraction path.

The width of RNP containment is defined in terms of the probability of the aircraft actually being where the navigation system defines its location. With an RNP of 1.0, there is a 95% probability that the airplane is within 1.0 nm (1.85 km) of either side of the intended flight path in the event of a "worst case" failure mode of the GPS satellite constellation. This failure mode involves an undetected failure of the most critical satellite at any given time. For RNP procedure design, this distance is doubled on each side of the path, resulting in a total containment width of 4.0 nm (7.41 km) for a RNP 1.0. The mathematical analysis for doubling the offset shows that there is a 99.99999% probability that the airplane is within this lateral containment.

The older, traditional ground-based navigation infrastructure of the global airspace system, developed in the 1940's, has two major components: a navigation system that relies on electronic signals emitted from ground-based transmitters; and an aircraft separation system based on a combination of ground-based radar and aircraft position reporting. This system has a number of major drawbacks, including geographic issues that limit the location and density of ground-based stations, particularly in mountainous terrain and across the earth's oceans, and a lack of precise positioning.

With RNP, the actual navigation position can be based on multiple inputs, including GPS, the onboard Inertial Navigation Systems, and the ground-based navigation systems. The Flight Management System (FMS) on board the aircraft blends this information to automatically provide positioning based on the accuracy and integrity of the received inputs. The high accuracy of this system allows flight path widths as narrow as 0.20 nm (0.37 km) or 2 x RNP 0.10 either side of the flight path centerline. In addition, constant angle vertical paths can be specified in the FMS such that procedures can be designed to provide three-dimensional lateral and vertical paths into and out of airports. These procedures allow the FMS to fly the procedure on autopilot, using the aircraft's automated systems.

The FMS is also able to use flight paths based on curvilinear paths, giving flight path designers considerable flexibility when compared to the traditional ground-based radio transmitters. This allows flight paths to be designed that go around obstacles or restricted airspace, have much smaller footprints, and allow landings with a low cloud cover. Figure 1 illustrates a RNP-based flight path. The Obstacle Clearance Area (OCA) for a RNP procedure is much smaller than the OCA for a traditional approach.

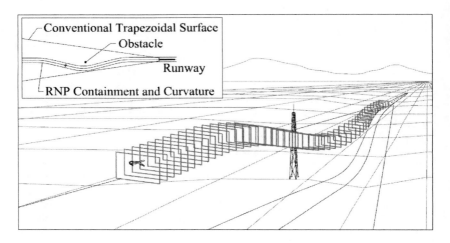

Figure 1. RNP Containment Illustration

When aircraft are monitored by radar controllers, the lateral separation standard between flight paths at the same altitude is 5 nm (9.3 km) in most instances. This is based on the inaccuracies of the aircraft's positioning system and on limitations inherent in ground-based radar systems. These systems provide a safe but inefficient use of airspace. The introduction of GPS position information into the aircraft flight management systems allows immediate and significant benefits in the design of terminal area approach and departure procedures, particularly in locations with limited ground-based infrastructure and mountainous terrain.

### The Need for Additional Capacity

Aviation agencies across the globe have been planning for growth in the air transport system. This growth can be accommodated in a number of ways, including construction of new runways or airports, use of new avionics and technologies, implementation of new air traffic management systems, or any combination of the above.

Airport throughput can be increased from the airside by reducing diversions and congestion and by increasing the efficiency of runway use. In addition to the terminal area, complex or congested airspaces between airports can be improved significantly. Clearly, throughput increases when efficiency increases, and efficiency increases when arrivals and departures are optimally spaced, and when weather does not diminish operations. RNP can remove or mitigate a number of these capacity constraints.

The key feature of an RNP procedure is the defined "tunnel" or containment that allows the airplane to curve precisely around terrain, obstacles, and restricted airspaces on both arrivals and departures. New or additional airport access, previously limited or unavailable by traditional rules, can be realized using RNP.

The Federal Aviation Administration (FAA) has been working on implementation of this new technology for a number of years (FAA 2006). With its current plans, the FAA expects: 1) increased safety; 2) improved airport and airspace access; 3) increased throughput; 4) increased schedule reliability; and 5) reduced delays. The FAA's strategy currently has three phases: 1) near-term (2006-2010), 2) mid-term (2011-2015), and 3) far-term (2016-2025). In the far-term, it is expected that RNP will be *mandated* in the busiest en-route and terminal airspaces. In conjunction with this strategy is the FAA's Next Generation Air Transportation System (NGATS), which encompasses goals that will integrate the national air transportation system with satellite-based navigation.

The International Air Transport Association (IATA) has also developed strategies for capacity enhancement through "globally harmonized air traffic management". In addition to expectations regarding access, capacity, and efficiency, they are considering environmental and security benefits, as well (IATA 2004).

How will this information and new technology affect airport planners? In some cases, RNP technology will open untapped opportunities to accommodate current and forecasted capacities.

Capacity can be increased when existing runway operations are optimized. RNP can enable simultaneous operations on parallel runways that are too close laterally for traditional operations. The FAA and MITRE are developing new capacity analysis models (FAA Air Traffic Organization 2006). RNP adds a new dimension to the way air transport capacity is evaluated. Aircraft can follow programmed tracks to avoid or mitigate impacts to noise sensitive areas, or avoid obstacles in areas that were previously unavailable due to the traditional, wide splays and clearance surfaces. RNP uses a much narrower and focused footprint.

Capacity can also increase by reorganizing and optimizing the use of busy airspace between nearby airports. Aircraft that follow RNP tracks can avoid conflicts or air space interactions with nearby airport operations. This has been evaluated for airspace interactions between JFK and LaGuardia airports and has demonstrated opportunities for significant annual savings (Dunlay and DeCota 2000). With a nationwide coordination of Air Traffic Control (ATC) operations, overall system capacity can be managed at the national level rather than from a patchwork of individually managed sectors. Aircraft separation standards can be based on the new technologies.

With respect to planning and budgeting, in some cases, it may be possible to delay or reduce the amount of new infrastructure that needs to be constructed. For locations with new airport and runway construction, the wide use of RNP will create options for a less restrictive runway orientation and location. Parallel runway configurations will require a smaller footprint. Orientation of the runway can be optimized to accommodate nearby airports or terrain features.

**The Benefits of RNP**

The benefits of an RNP-based airspace infrastructure can be described in three broad categories when compared with the traditional airspace infrastructure: 1) increased total system throughput; 2) reduced fuel consumption between city pairs; and 3) increased total system safety.

Increased system capacity, or throughput, is achieved through several mechanisms, including increased capability to arrive and depart from airports in inclement weather conditions, increased payload per aircraft operation, and increased airspace utilization through the use of RNP separation standards. When designing instrument procedures based on the traditional ground-based infrastructure, procedure designers have to use OCAs that can be several miles across and are constrained to straight line paths. In many cases, these large OCAs and straight line paths result in high minimum altitudes and less than optimal aircraft positioning in order to avoid obstacles. These high altitudes and sub-optimum positions often become problems when designing procedures into and out of airports.

On an approach procedure with cloud cover, aircraft are not allowed to descend below the specified Minimum Descent Altitude (MDA) to land, unless the pilot can see the runway or surrounding environment. In mountainous regions, straight line paths encounter obstacles thousands of feet above the airport elevation, resulting in MDAs that are higher than prevailing cloud ceilings, severely restricting throughput in all but the most ideal weather conditions. In many instances, flight paths must not only avoid obstacles, but they must avoid certain restricted airspace, such as other airports, other approach and departure paths, military installations, wildlife preserves, and noise sensitive areas. RNP procedures provide positive lateral and vertical guidance along the entire approach path to the runway. As a result, operators with RNP approaches have an access advantage over operators that must use the traditional navigation systems.

As in approach procedures, traditional instrument departure procedures often have high minimum altitudes on segments close to the airport due to the large number of obstacles and terrain features encountered within the departure procedure OCA. Since the aircraft's ability to climb is dependent on its weight, aircraft operators must ensure that the weight of the airplane at takeoff is such that the required departure altitudes can be attained. In mountainous terrain, takeoff weights are often severely restricted due to the high segment altitudes that must be achieved. These restricted takeoff weights are achieved by limiting the payload (passengers, cargo, and fuel) that the aircraft can carry, representing a decrease in system throughput. RNP procedures often allow aircraft to depart with higher payloads in inclement weather conditions, providing positive lateral and vertical guidance along the entire departure path.

Fuel savings and the resultant reduction in greenhouse gasses are achieved through several means, including a reduction in total distance between city pairs, direct flight paths rather than vector maneuvers, the ability to descend along constant descent paths at reduced power settings during approach procedures, and a reduction in the number of airborne system disruptions due to inclement weather, such as diversions to alternate airports or holding above destination airports. IATA's Save One Minute campaign maintains that a reduction of one minute from every flight across the globe would save about $4 billion annually, based on the average airline operating cost of $100 per minute (Paylor 2005).

Some types of traditional approach procedures require the aircraft to "step down", flying a series of descending and level segments in the landing configuration, with flaps and landing gear extended. These level segments at low altitude in the landing configuration require high power settings and a resultant increase in fuel consumption and noise. With a RNP approach (constant descent angle), a typical approach procedure would consume less fuel, up to 112 gallons (420 l), than a traditional procedure to the same runway (Airservices Australia 2005). For a major airport with 90 arrivals per hour, this is a potential savings of 10,000 gallons (38,000 l) per hour and a reduction of about 100 pounds (45 kg) in carbon-dioxide generated per hour.

One of the largest threats to safety in the global air transportation industry is Controlled Flight Into Terrain (CFIT) accidents. The majority of these accidents occur when aircraft inadvertently descend too low flying an instrument approach, and subsequently impact the ground short of the runway. The primary cause of these accidents is a loss of situational awareness on the part of the flight crew. Many traditional procedures require pilots to identify the aircraft's lateral and vertical position based on multiple facilities and multiple gauges in the cockpit. These types of procedures are known as non-precision approaches and represent the majority of the approach types found in areas with limited infrastructure. Instrument approach procedures that provide a constant angle descent to the runway have been determined to be five times safer than non-precision approaches (Flight Safety Digest 1996).

RNP procedures can also include pre-programmed track guidance during emergency conditions such as an engine failure. The engine-out track follows a path and reduced climb gradient that is known to be clear of all obstacles. Once the pilot selects the engine-out procedure from FMC, all attention can be focused on the failed engine and the ensuing tower communications.

The first airline to realize the benefits of RNP was Alaska Airlines, at Juneau International Airport in the state of Alaska, in 1994. Juneau, a challenging location surrounded by mountains, is an important hub operation for Alaska Airlines, which operates over 20 Boeing 737 flights a day into and out of the airport. The traditional instrument approach procedures did not provide true all-weather capability and were poorly aligned with the landing runways. As a result, Alaska Airlines' pre-RNP operations into Juneau were often halted for days due to inclement weather conditions, resulting in economic losses to the airline and the local community. Following development and obtaining the required approvals, the RNP procedures represented a significant change in Juneau for Alaska Airlines, resulting in a savings of several million dollars each year in the Juneau operation alone.

Another early user of RNP is WestJet, based in Calgary, Alberta, Canada. WestJet already has RNP procedures to most of its Canadian airports. It has been estimated that using a single RNP approach at Kelowna, B.C. saved 15 nm (28 km) on each flight from Calgary, translating into a total of $1.5 million (Canada) a year. That figure relates only to fuel and time savings and takes no account of the additional benefits achieved from fewer weather-related diversions (Paylor 2006).

**Example Use and Benefits of RNP Geometry**

The following example illustrates how a narrow, fixed-width containment, that follows programmed curves and tangents, and uses the aircraft's specific performance characteristics, can create new opportunities for improved access and capacity at Portland International Airport (PDX) in Oregon.

Currently, 99% of PDX's operations occur on its two main runways, which are parallel and staggered. Runways 10L and 10R are approached from the west and runways 28L and 28R are approached from the east. Runway 21, which intersects the main runways, is currently used only when major storms arrive from the southwest, or for about 1% of all current PDX arrivals (Hanson 2007). This seldom-used runway is not necessarily a high-priority candidate for RNP procedures, but it does demonstrate a number of RNP benefits in the following example.

It should be noted that Alaska Airlines and Horizon Air currently have RNP procedures for runway 28L and 28R arrivals, as part of an early FAA RNP implementation program.

Strong southwest winds occasionally create crosswind limitations for runways 10 and 28. As a result, controllers may offer RW21 for alternate arrivals. In some cases, all approaches are required to use RW21. See Figure 2.

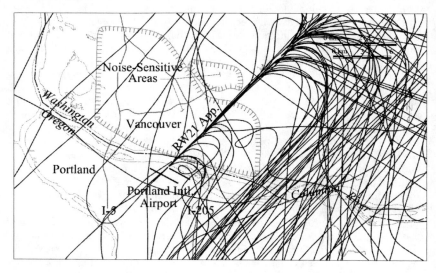

Figure 2.  Actual Approach Tracks to PDX RW21 on 14-Dec-2006

The current approach procedure for RW21 is based on a ground-based localizer signal which is "captured" along a straight line approach path, up to 17.5 nm (32.4 km) from the runway. Since the localizer provides no vertical guidance, the approach consists of a series of undesirable altitude step downs. Furthermore, residential developments in the City of Vancouver, on the Washington side of the Columbia River, have become noise-sensitive areas.

A 3-degree glide path angle was selected for the example approach.  The resulting Vertical Error Budget (VEB) of 285 feet (86.9 m) provides ample vertical clearance in the final segment between the Final Approach Fix (FAF) and RW21. The missed approach track (not shown) would mimic the non-RNP missed approach track in use today, which consists of a right turn back to the north.  In this example, if RNP approaches to RW21 are provided as shown in Figure 3, the benefits listed below could be realized:

- Reduced track mileage.    Typically, arriving airplanes intercept the localizer's signal about 10 nm (18.5 km) from the runway.  With the RNP approach, the airplane is on the extended runway centerline 6 nm (11 km) from the runway, saving about 4 nm (7.4 km).  Airplanes arriving from the south or east have significantly reduced track mileage.

- Increased access, safety, and capacity during inclement weather (fewer diversions or delays).  During periods with only RW21 operations, RNP would enable an efficient throughput.

- Repeatable tracks that minimize impacts to noise-sensitive areas.    The arrival tracks during current operations are scattered widely.  A plot of the arrival tracks using RNP would show all of the tracks directly on top of the three designed flight paths shown in Figure 3.

- Should either of the two main runways be taken out of service for maintenance activities, RNP procedures for RW21 would be available.

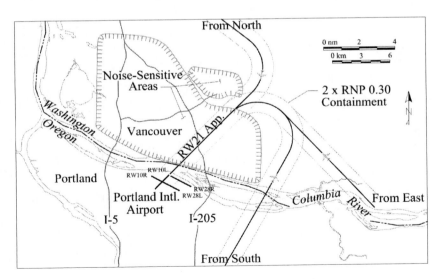

Figure 3.  Example of New RNP Approaches to PDX RW21

**Conclusions**

Benefits can be realized at existing airports in developed areas as well as at new facilities. Also, runway approaches with undesirable procedures can be, and should be, improved. Reduced track mileage and fewer diversions save time and fuel, and generate fewer emissions. For example, assuming that an average reduction of 4 nm (7.4 km) could be achieved for each of Portland's procedures, and assuming 100,000 airline operations per year, a 400,000 nm (740,000 km) reduction in total track mileage would be achieved. This would yield an annual savings of about 1.5 million gallons (5.7 million l) of fuel, about $13 million in total airline operating costs, and about 16,600 tons (15,100 metric tons) of carbon-dioxide generated.

Additionally, a constant descent profile managed by the onboard FMS, would result in lower power settings for the engines, which would use less fuel and produce less noise. Operators with older fleets will require equipment upgrades before they can fully realize the benefits of RNP. The opportunities that RNP offers are vast and challenging, and regulatory agencies across the globe are faced with the task of integrating and supporting new criteria and standards associated with RNP. In the regulatory and standards world, these changes require working through comprehensive processes, involving the coordinated activities of the respective stakeholders. The list of stakeholders includes: regulatory agencies, air traffic services, airport operators, aircraft operators, industry manufacturers, environmental groups, and airport planners and designers.

The rules are changing and the Third Revolution of Flight is underway. With RNP curvilinear paths, civil engineers can now apply their transportation and geometric alignment expertise to aircraft route design. For improved safety, efficiency, access, and capacity, implementing RNP on a global scale is the right thing to do.

**References**

Airservices Australia (2005). "Airservices Australia Fuel Saving Initiatives From the Ground Up". Waypoint 2005. Sydney, Australia.

Dunlay, William Jr. and William DeCota (2000). "Flight Path Precision." Civil Engineering - Volume 70 No. 12, Reston, VA. 56-61.

Federal Aviation Administration (FAA) Air Traffic Organization (2006). "MITRE Researchers Showcase Aviation-Related Projects." ATO Online, Washington, D.C.

Federal Aviation Administration (2006). "Roadmap for Performance-Based Navigation Version 2.0". Washington, D.C.

Flight Safety Digest (1996). "Airport Safety: A Study of Accidents and Available Approach-and-landing Aids." Volume 15 No. 3, Alexandria, VA.

Hanson, Eric (2007). Personal Communication. January 18, 2007, Portland, OR. (FAA Terminal Operations, Western Service Area, Portland Tower/TRACON.)

International Air Transport Association (2004). "One sky... ...global ATM – ATM Implementation Roadmap Version 1.0." Montreal, Quebec.

Paylor, Ann (2005). "Waste Not, Want Not." Air Transport World, Brussels, Belgium, p. 64.

Paylor, Ann (2006). "Tunnels in the Sky." Air Transport World, Brussels, Belgium, p. 57.

# Integrated Model for Air Traffic Control, Aviation Weather and Communication Systems

Chuanwen Quan,[1] Antonio A. Trani,[2]

[1]PB Americas, Inc., 312 ELM Street, Suite 2500, Cincinnati, OH 45202; PH: (513) 639-2174; Fax: (513) 421-9657;
email: quanc@pbworld.com
[2]Civil and Environmental Engineering Department, Virginia Tech, Blacksburg, VA 24061; PH: (540)231-4418;
email: vuela@vt.edu

## Abstract

In this paper, a Systems Dynamics methodology to model and simulate air traffic, aviation weather and communication systems is presented. Two computer models are developed with ITHINK and MATLAB software packages: 1) Air Traffic Flow Model (ATFM) and 2) Aviation Weather Information and Bandwidth Requirement Model (AWINBRM). The ATFM model is used for quantifying the volume of air traffic at each phase of flight in three flight regions. The results of the ATFM model are used as inputs for the AWINBRM model. The AWINBRM model is used to define required aviation weather information and their sizes at each phase of flight in three flight regions. These models could be deployed as decision support systems to aid airport and airline personnel. The models' results demonstrate that all types of aircraft need large quantities of weather data before taking off from the airport. The aircraft in the terminal area require less aviation weather data but the weather may have the most impact in terms of flight safety. Models also provide detailed information for planning aviation weather communication systems in the future.

## Introduction

### Air Transportation

The demand for air transportation services, measured in terms of total passenger enplanements at more than four hundred U.S. airports with commercial service, tripled between the years 1976 and 2005 according to the FAA's Terminal Area Forecast (TAF 2006). The number of annual air carrier commercial flights at all U.S. commercial airports increased 42% in the same period from 9.5 to 13.5 million. Similarly, the number of commuter and air taxi flights tripled during the same period. The events of September 11, 2001impacted negatively the US air transportation industry, when the

132

average seating capacity of the U.S. carriers declined 15 percent. However, due to lower fares and a solid economic recovery, it is reported that airlines have flown the number of enplanements recorded in 2005. In the future, the TAF projects that the number of passenger enplanements and the number of flight operations at U.S. commercial airports are expected to increase at annual compound growth rate of 3.1 and 1.6 percent from 2005 to 2025, respectively. The growth of commercial passengers and flights traversing in the National Airspace System (NAS) has outpaced the growth of NAS infrastructure.

Historically, the FAA has spent billions of dollars to modernize the NAS through the development, acquisition, implementation of new ATC/ATM technologies, and airport expansion. Many of these capabilities entrenched in the FAA programs have made claims that delays will be reduced significantly as air traffic demand grows. However, some recent developments indicate that air transportation delays are increasing at a faster pace than justified by the demand for air transportation service. In 2000, one in every four flights was delayed, canceled or diverted according to FAA statistics. The equivalent of 163 million fliers was delayed according to data compiled by the Department of Transportation Inspector General. Aircraft departure and arrival delays increased 126% between 1997 and 2000. Based on work done by Russ Chew at ATA, if the air traffic control system is not fixed and if the events of September 11, 2001 would not have occurred, ATA estimates that delays would have increase by some 250% by 2005. Although this situation would not occur due to the impact of September 11 terrorist attacks, the flight delays have started to creed up after the spring 2004 as air traffic reaches or exceeds the record levels experienced in the year 2000. Based on current statistical data from the Bureau of Transportation Statistics, 22 percent of flights on large airlines were delayed at least 15 minutes and aircraft were delayed 22.1 million minutes—a record high in 2006. Before September 11, 2001, air traffic delays were costing airlines and their passengers an estimated $6 billion a year. Today, individual airlines say delay costs are mounting.

### Causal Factor

Aviation efficiency and safety are affected by a number of factors such as traffic volume, airside conditions, aircraft, and weather etc. The weather system is a major contributor to aviation delays, congestion, accidents, and incidents, because air traffic and weather systems sometimes occupy the same airspace and uncertainty exists about the weather. In the past five years, weather systems have accounted for 71% of all delays in the NAS. According to statistics compiled and published by the FAA, more than 40% percent of the GA accidents are attributed to weather. Today, there are many aviation weather products and information packages available in the aviation industry. Unfortunately, the lack of efficient dissemination of weather data products to pilots and the frequencies used to obtain aviation weather information often become saturated and at times limiting access to the information when it is needed most. Pilots can tune radios to receive automated weather services, for example, Hazardous In-flight Weather Advisory Service (HIWAS) and Automated Terminal Information Service (ATIS) etc. However, the information from these aural sources is limited and may be relevant only for very localized area. But there may be another reason.

Both analytical and computer models are critical tools to provide analytical research capability for the NAS. If we wish to conduct tests and experiments, or deploy new equipment or procedures, we cannot simply halt NAS operations—The NAS operates constantly and continuously. Nor can we simply plug in advanced prototype systems for testing during NAS operations. Models are simply the easiest and more justified tool to apply.

## Methodology and Approach

### *Objectives*

In this research, two computer models are developed. The models are used to quantify air traffic flow volume at each phase of flight or in three flight regions: en-route, in the airport terminal airspace, and on the ground, to determine aviation weather information requirements at each phase of flight or region, and to quantify their bandwidth requirements. Furthermore, the results from those models can be used to determine various alternatives for future aviation communication systems.

The ultimate goal of the models described here is to serve as a laboratory where policies can be experimented with and tried before implementing them into the real world—the NAS system. Moreover, these computer models can evolve dynamically through time, and as operations dictate, allowing decision makers to exercise policies at various points in time to quantify results cost-effectively.

### *Methodology*

The Systems Dynamics Concept adopted in this research is used to capture decisions as they take place in an information feedback system. Systems Dynamics is based upon: (1) decision-making, (2) feedback systems analysis, and (3) simulation. Decision-making states how action is to be taken. Feedback deals with the way information is to be used for decision-making. Simulation permits decision makers to view the implications of their decisions.

Two software packages will be used in developing this research: "ITHINK", and "MATLAB". "ITHINK" is the Systems Dynamics software— Through systems thinking, it offers a better conceptual framework for underlying construction and subsequent simulation of mental models. "MATLAB" is a high-performance language for technical computing—It integrates computation, visualization, and programming in an easy-to-use environment where problems and solutions are expressed in a familiar mathematical notation.

### *Approach*

Based on the above methodology, models will be designed to analyze an airport system, an Air Route Traffic Control Center (ARTCC) or an entire country. They will be used as a resource planning tool to formulate short – to long-term improvements in the NAS system.
Two models are developed in this research:
- Air Traffic Flow Model
- Aviation Weather Information and Bandwidth Requirements Model

Air Traffic Flow Model (ATFM) is based on the principles of Systems Dynamics and continuous simulation modeling to model aggregate traffic flows. ATFM uses the premise that aircraft inside a region of airspace (an Air Route Traffic Control

Center – ARTCC, for example) can be in any one of flight phases which will be discussed in the following section.

Each phase has specific aviation weather information and communication requirements. The dwell time for each state has been derived from actual flow analyses using FAA Enhanced Traffic Management System (ETMS) data, BADA (Eurocontrol aircraft performance data), Official Airline Guide (OAG), Flight Information Display System (FIDS), Terminal Area Forecast (TAF), simulation case analysis, and observation data at several airports etc. For the sake of discussion in this paper, the data used in the modeling are representative of traffic flows of the ARTCC. However, the model can be tailored to represent any region of airspace desired with minor modifications to some model parameters. Based on the aircraft characteristics, four sub-air traffic flow models have been developed using the same principles but different aircraft parameters such as dwell times and speed etc., which are air carrier, air taxi/commuter, general aviation (GA), and military models.

Using the traffic volume at each phase of flight output from the ATFM as input, the Aviation Weather Information and Bandwidth Requirements Model (AWINBRM) is developed. There are nine aviation weather products used for making the tactical decision and fourteen strategic aviation weather products. Each aviation weather product has been modeled for each aircraft category. Usually each aviation product has its own distribution rate and life-time. The AWINBRM model maintains the validation of the aviation weather products and updates the aviation weather products immediately as soon as the updated aviation weather products are published. Based on the results from the AWINBRM, the existing and potential communication systems used for transmitting aviation weather information are explored in this research.

## Model Development

### Air Traffic Flow Model

#### Flight Segment and Phase

Typically, a flight takes off from its original airport, cruises in the airspace, and then lands at its destination airport. In reality, this flight has many flight phases and segments and needs various types of information to operate safely, including weather data, at each phase. In this research, the totality of a flight operation is divided into three distinct segments - Departure Segment, En-route Segment, and Arrival Segment and twelve phases of flight including an alternative operation. The flight segments and phases are listed in Table 1.

**Table 1: Flight Segment and Phase**

| Flight Segment | Flight Phase |
|---|---|
| Departure Segment | Pre-Planning; Pre-Operation; Taxi-out and Take-off Operation; and Departure Operation |
| En-route Segment | Initial Climb Operation; Initial Cruise Operation; and Cruise Operations |
| Arrival Segment | Approach Operation; Landing Operation; Taxi-in and Parking Operation; Post Flight Operation; and Alternative Operations |

*Model Causal Diagram*

A general causal diagram depicting the possible transition flights inside an ARTCC Center is shown in **Figure 1**. The diagram distinguishes between information flows (dashed arrows) and accumulation flows (solid arrows) to showcase fourteen aircraft states (boldfaced names) inside the volume of airspace of interest. Causal diagrams are standard techniques used in Systems Dynamics modeling. The polarity of the causal relationships shown in the diagram represents the general trend of the slope relating two immediate variables. Inbound Aircraft Demand Function (IADF) "causes" an increase in the Number of Inbound Aircraft Entering Cruise Operation (NIAEC), and in the Number of Transient Aircraft Entering the Center Airspace (NTAECA). The left part of the figure describes inbound aircraft operations entering the center cruise operations (NIAC), following approach operations (NIAA), landing operations (NIAL), taxi-in and parking operations (NIATIP), to post flight operations (NIAP) and aircraft idle condition (IAA). The right part of the figure shows outbound aircraft operations beginning preflight and flight plan filing (NOAPP), continue to preflight operations (NOAP), taxi-out and take off operations (NOATOTO), departure operations (NOAD), initial climb segment operations (NOAICS), initial cruise operations (NOAIC), and up to enter cruise operations (NOAC). The plus and minus signs represent air traffic flow changes - increase and decrease, respectively. **Figure 2** depicts this causal diagram for the typical air carrier aircraft category.

*Model Equation*

Each state and rate variable in the models has its equation. The following is an example for the state variable of the Number of Inbound Aircraft in Approach Operation (NIAA). NIAA is dependent on the value of the NIAA in previous time, the Number of Inbound Aircraft Leaving Cruise Operation to Approach Operation (NIALCA) rate, and Number of Inbound Aircraft Leaving Approach Operation to Landing Operation (NIALAL) rate. The equation for the NIAA is expressed mathematically as:

$$\frac{d}{dt}(NIAA_t) = \left( \frac{d}{dt}(NIAA_{t-1}) + (NIALCA_t - NIALAL_t) \right)$$

*Graphical User Interface*

To simplify the interaction between a user and the model a simple Graphical User Interface (GUI) was developed for the ATFM. GUIs are standard features in the modeling approach adopted in ITHINK. The GUI for each aircraft category has been developed. The model interfaces include input parameters, output results including tables and graphics, and control buttons which can be used to change and modify input parameters, re-run, stop, pause models or go to any other models. **Figure 3** illustrates an example of the interface for air carrier traffic flow model.

## Aviation Weather Information and Bandwidth Requirements Model

*Aviation Weather Domain*

The industry generally recognizes two kinds of airborne weather-related decisions, "tactical" and "strategic", which could be interpreted simply as "now" and "near future".

"Tactical" decisions are essentially reactive flight trajectory decisions which need to be made quickly with whatever information is at hand. Strategic decisions, on the other hand, tend to be more pro-active, planning for avoidance rather than weather penetration. These decisions are characterized by the ability to identify a hazard early, collaborate on a plan to avoid it, and make relatively small, well-coordinated changes to the flight trajectory.

*Aviation Weather Products*

For decision-making purposes, aviation weather products are divided into two categories: tactical and strategic aviation weather products. For making tactical decisions, there are nine types of aviation weather products which including observation data and forecasts can be used. **Table 2** lists all tactical aviation weather products and their properties.

**Table 2: Tactical Aviation Weather Products and Their Properties**

| Products | Application Area | Types | Product Rate (times/day) | Product Life | Estimated Size (byte) |
|---|---|---|---|---|---|
| METAR/SPECI | Terminal Area | Current | 24 | 1 | 500 - 1,000 |
| TAF | Terminal Area | Forecast | 4 | 24 | 500 - 1,000 |
| Area Forecast | En-route and Terminal Area | Current and Forecast | 3 | 8 | 3,000 - 10,000 |
| Route Forecast | Air Traffic Routes | Forecast | 3 | 15 | 500 - 2000 |
| AIRMET - Sierra | En-route and Terminal Area | Current and Forecast | 4 | 6 | 500 - 1,000 |
| AIRMET - Tango | En-route and Terminal Area | Current and Forecast | 4 | 6 | 500 - 2,000 |
| AIRMET - Zulu | En-route and Terminal Area | Current and Forecast | 4 | 6 | 500 - 2,000 |
| Domestic SIGMET | En-route and Terminal Area | Current and Forecast | 6 | 4 | 500 - 1,000 |
| Convective SIGMET | En-route and Terminal Area | Current and Forecast | 6 | 4 | 1,000 - 5,000 |
| International SIGMET | En-route | Current and Forecast | 6 | 4 | 500 - 2,000 |
| Winds and Temperature Aloft | En-route and Terminal Area | Forecast | 2 | 6/12/24 | 250 - 500 |
| Severe Weather Forecast Alert | En-route and Terminal Area | Forecast | as required | 1 | 250 - 500 |
| Center Weather Advisory | En-route and Terminal Area | Forecast | as required | 2 | 500 - 1000 |
| PIREP Distributed | En-route | Current | as required | 1 | 250 - 500 |

There are fourteen strategic weather products which are designed to support pre-flight planning and strategic decisions. Their applicability to tactical decision making is limited due to the update rates which vary from 1 hour to 12 hours. **Table 3** lists all strategic aviation weather products and their properties.

*Model Flow Chart and Equation*

**Figure 4** depicts graphically the flow chart for Aviation Weather Information and Bandwidth Requirement Model (AWINBRM) including two major parts:
(1) Uplink component, which means that the aviation weather information is transmitted from ground service centers such as National Weather Service (NWS) to the aircraft. This includes the tactical and strategic aviation weather product sections. In the tactical aviation weather section, aviation weather products are mainly considered the observation data (current condition) and the nowcast weather products that usually are one to three hours forecasts. Four sub-tactical aviation weather product models are developed. They are air carrier, air taxi/commuter, general aviation, and military tactical

**Table 3: Strategic Aviation Weather Products and Their Properties**

| Products | Application Area | Types | Product Rate (times/day) | Product Life | Estimated Size (kb) |
|---|---|---|---|---|---|
| Radiosonde Additional Data | En-route and Terminal | Current | 2 | 12 | 2 -50 |
| Constant Pressure Analysis Charts | En-route and Terminal | Current | 2 | 12 | 50 -500 |
| Composite Moisture Stability Chart | En-route and Terminal | Current | 2 | 12 | 50 - 500 |
| Low Level Significant Weather Program | En-route and Terminal | Forecast | 4 | 12/24 | 20 - 200 |
| High Level Significant Weather Program | En-route and Terminal | Forecast | 4 | 6 | 20 - 200 |
| Convective Outlook | En-route and Terminal | Forecast | 5 | 24/48 | 5 - 15 |
| Surface Analysis Charts | En-route and Terminal | Current | 8 | 3 | 100 - 1000 |
| Weather Depiction Chart | En-route and Terminal | Current | 8 | 3 | 30 - 300 |
| Radar Weather Report | En-route and Terminal | Current | 24 | 1 | 800 - 2000 |
| Radar Summary Chart | En-route and Terminal | Current | 24 | 1 | 300 - 3000 |
| Satellite Weather Pictures | En-route and Terminal | Current | 72 | n/a | 1000 - 5000 |
| Hurricane Advisory | En-route and Terminal | Advisory | 300 nm offshore | n/a | 5 - 50 |
| Meteorological Impact Statement | En-route and Terminal | Forecast | as required | 2/12 | 3 - 30 |
| Volcanic Ash Forecast and Dispersion Chart | En-route and Terminal | Forecast | 4 | 6/12 | 6 - 60 |

aviation product models. Similar to the tactical aviation weather product model, the strategic aviation weather product section also has four sub-models for air carrier, air taxi and commuter, general aviation, and military aircraft categories, respectively. The strategic aviation weather products are focused on some nowcasts and the short and longer forecasts aviation weather products.

(2) Downlink component, which means that the aviation weather data observed from the sensors on board are transferred to data processing center on the ground or to the aircraft directly.

For future development, the AWINBRM would also include the aviation weather improvements such as update weather product sampling rate, coverage area, uncoded, and graphic weather product, etc.

The causal diagram for AWINBRM is depicted in **Figure 5**. The important factor for determining weather product requirement and bandwidth is how many times a flight will request for one aviation weather product at each phase of flight. This is called the aviation weather product sampling times (WPST), which is largely dependent on the dwell time from the outbound initial phase of flight (preflight planning and flight plan filing) or inbound initial phase of flight (cruise operation) to the phase of flight that is determined, weather sampling rate, weather sampling times in the previous phases of flight, and weather product distribution schedules. The following equation is used to determine the weather product sampling times at outbound taxi-out and take-off phase of flight. Other equations in the AWINBRM model for determining weather product sampling times at different phase of flight are similar to this equation.

WPSTOTOTO = IF((INT(DTOPPTOTO/Weather Sampling Rate)- Weather Sampling Times from PP to P)>=1 or METAR Time Factor=1)
THEN (1)
ELSE (0)
Where:

WPSTOTOTO = Weather product sampling times at taxi-out and take off phase (time)

DTOPPTOTO = The total dwell time from Preflight Planning to Taxi Out and Take Off for outbound aircraft (minute)

Weather Sampling Rate = Weather product distribution frequency per day (once/minute)

Weather Sampling Times from PP to P = The total sampling times from Preflight Planning to Preflight for outbound aircraft (time)

Weather Time Factor = Weather product distribution schedule (0 or 1)

INT = Determine integer

*Graphical User Interface*

Like the ATFM, nine model graphical user interfaces for the AWINBRM are developed. Eight of them express tactical and strategic weather product for each aircraft category, and one reflects the aircraft as the sensor interface. **Figure 6** demonstrates the interface of the AWINBRM for air carrier example.

**Model Application**

The Atlanta Air Route Traffic Control Center (ARTCC) was selected as the environment to run the ATFM and AWINBRM models within. The following are the partial results from the application of these models in Atlanta ARTCC.

**Figure 7** demonstrates the changes of the air carrier traffic at the airport, in the terminal, and in the en-route airspace in Atlanta ARTCC over a 24-hour period. There are two peak operations for air-carrier traffic, which occur during the morning and evening peak period. The morning peak aircraft operations are mainly comprised of departure flights, while most of the arrival flights occur in the evening peak. Most air carrier aircraft are at the airport because air carrier pilots need to take longer times to collect data than other aircraft categories. Also air carrier aircraft need more time to load and unload passengers at the airport. The number of air carrier aircraft in the terminal area during this period is minimal, because typically, terminal areas are relatively narrow and the separations between airplanes have to always be accurately maintained. The maximum number of aircraft at the airport reaches about 205 during the morning and evening. The maximum number of aircraft in the terminal and in the airspace is 135 and 155, respectively. As shown in the figure, most air carrier flights operate between 6:00 A.M. and 22:00 P.M.

The tactical aviation weather information requirements at the airport, in the terminal, and in en-route airspace operations over a 24-hour period for air carrier aircraft in 2005 is as shown **Figure 8**. The tactical aviation weather product requirements over the time in three regions vary with the changes of air carrier traffic. The more aircraft operations in the region, the more the tactical aviation weather product would be needed.

The tactical aviation weather information requirements shown in the figure for the operations at the airport are larger than the requirements in the terminal or in the airspace. This is because more airplanes are at the airport, where pilots need to collect more tactical aviation weather information before planning a flight. Maximum tactical aviation weather product requirement ranges around 75 MB for the operations at the airport. The tactical aviation weather information requirement in the airspace is the second largest in three regions. The lowest tactical aviation weather information requirement occurs in the terminal area. These results are consistent with the traffic volumes generated from the ATFM model.

The variations in the strategic aviation weather information requirements at the airport, in the terminal, and in the airspace over a 24-hour period for the air carrier aircraft category in 2005 is depicted in **Figure 9**. There are two pronounced peaks for the strategic aviation weather information requirements in the airspace, and one peak at the airport for the air carrier aircraft category. Maximum strategic aviation weather information requirement for air carrier aircraft at the airport is approximately 15 GB. This requirement is larger than that in the airspace or in the terminal; because pilots typically collect more strategic aviation weather information for origin and destination airports and en-route before their flight take-off. Maximum strategic aviation weather information requirements for air carrier aircraft in the airspace is approximately 2.3 GB. The sawtooth pattern shown in this figure seems to indicate that the interval for updating strategic aviation weather products is potentially larger than the tactical aviation weather products.

Variations in the icy conditions, turbulence, wind aloft and temperature, and convective weather data for a 24-hour period in the terminal area are depicted in **Figure 10**. The size of the wind aloft and temperature data is the largest in the four weather data. Maximum data requirements for the wind aloft and temperature is approximately 0.9 MB

The model also can run the future years. **Figure 11** illustrates the variations in the tactical aviation weather information requirements over a 24-hour period for air carrier category at the airport, in the terminal, and in the airspace in 2025. As previously described, there are also two pronounced peaks for tactical aviation weather information requirements in the airspace operations, and one at the airport.

Through exploring the existing aviation weather information communication systems from aviation datalink systems to voice systems, it becomes clear that these systems can not possibly meet all the requirements needed, especially, in the peak operation period. In the future, aviation industry would need to explore other potential solutions to solve this problem, such as Microwave Distribution Systems and Internet In/From the Sky etc.

**Conclusions**

This paper demonstrated through the results from the Air Traffic Flow Model (ATFM) application, that all types of aircraft invariably get delayed on ground at airports, followed by the en-route airspace, and the least aircraft delays occur in the terminal airspace. The findings from the AWINRM seem to support the following:

1) In the preflight operations, pilots (and/or dispatcher and ATC) may need large quantities of data from a number of sources to obtain an accurate picture of the current and forecasted conditions along the intended route.

2) Aircraft operations in the terminal area need less aviation weather information than has been previously identified as the area where weather may have the most impact in terms of flight safety.

3) Aviation industry needs to provide both strategic and tactical (for forecasted and current conditions) information as a function of flight phase, to increase the updated rate associated with the number of weather products used in pre-flight planning, and to add on-board sensors to address in-flight icing and turbulence.

Except for enhancing the existing aviation weather information communication systems, the aviation industry should explore other potential communication systems to meet the future expanded requirements.

## References

Air Traffic Weather Requirements Team, (1993). *Air Traffic Weather Requirements Report*.

ATAC Corporation, (2001). *SIMMOD PLUS Users Manual*, Sunnyvale, CA.

Ball, J.W. et al., (2000). *Aviation Weather Information Communications Study (AWIN)* Phases I and II, Lockheed Martin Aeronautics Company, Marietta, Georgia.

Eurocontrol, (1998). Aircraft Performance Characteristics

High Performance System, Inc., (2001). *ITHINK 7.0 Users Manual*, Hanover, NH.

http://www.air-transport.org

http://www.faa.gov

http://www.usatoday.com

Keel, Byron M. et al., (2000). *Aviation Weather Information Requirements Study*, Georgia Tech Research Institute, Atlanta, Georgia.

Louisville International Airport, (2003). *Flight Information Display System*.
Mathworks Inc., (1997). *Matlab 5.3 Users Manual*, Natick, MA.
Official Airline Guide, (2003).

PB Aviation, Inc., (2001). *Sacramento International Airport Master Plan Update Study*, Cincinnati, OH.
PB Aviation, Inc., (2001). *Pittsburgh International Airport Master Plan Update Study*, Cincinnati, OH.

The Preston Group, (1998). *TAAM Users Manual*, Fairfax, VA.

Trani, A.A., et al., (1998). *Integration of Reusable Launch Vehicles into Air Traffic Management*, NEXTOR Research Report RR-98-15.

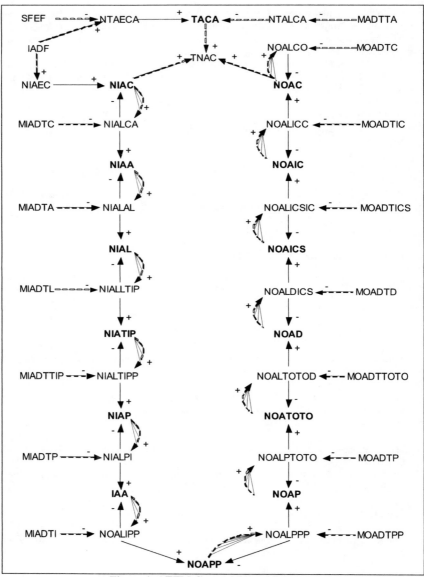

**Figure 1. ATFM General Causal Diagram**

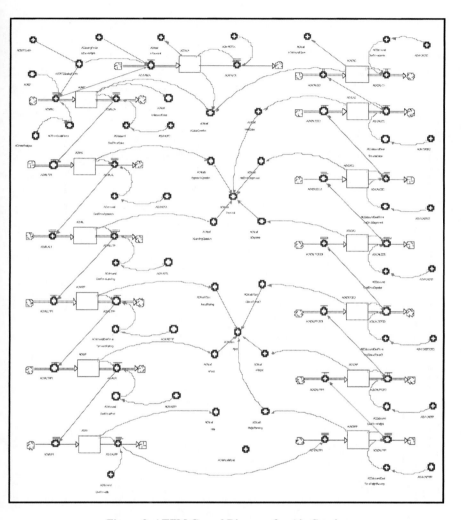

**Figure 2. ATFM Causal Diagram for Air Carrier**

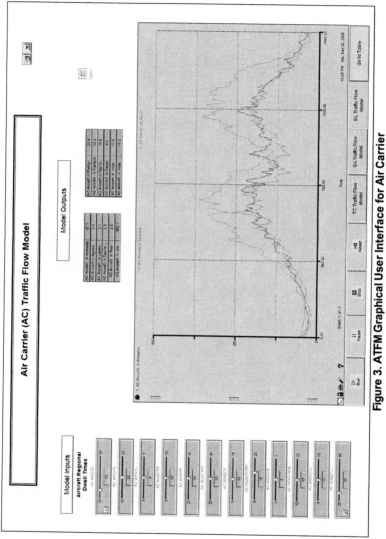

**Figure 3. ATFM Graphical User Interface for Air Carrier**

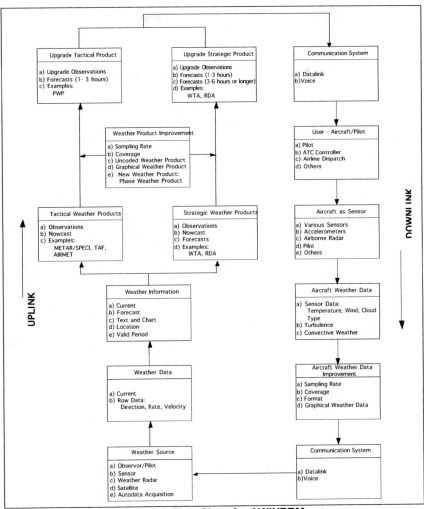

**Figure 4. Flow Chart for AWINBRM**

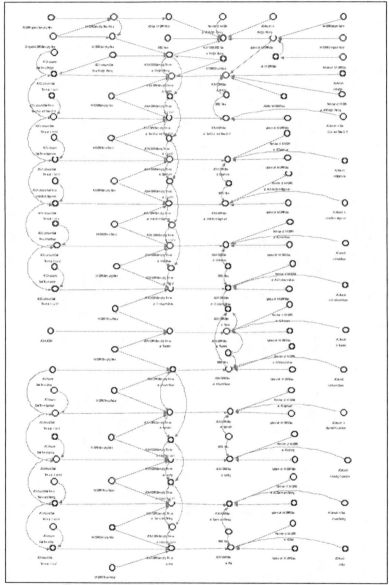

**Figure 5. AWINBRM Causal Diagram for Air Carrier**

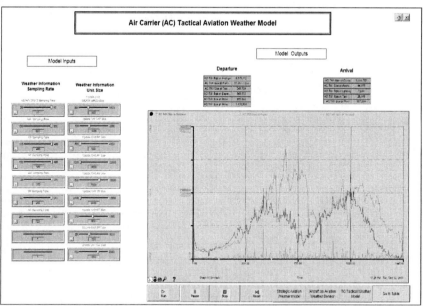

**Figure 6. AWINBRM Graphical User Interface for Air Carrier**

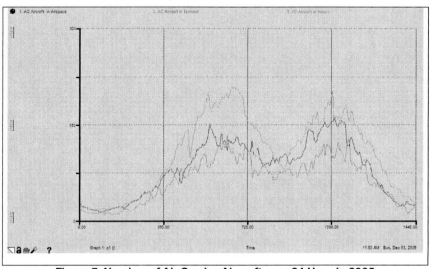

**Figure 7. Number of Air Carrier Aircraft over 24 Hour in 2005**

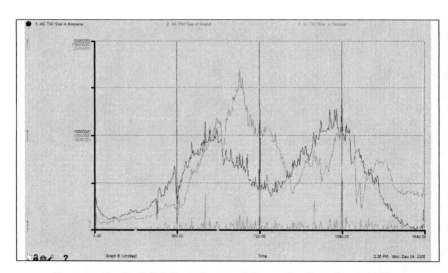

**Figure 8. Tactical Aviation Weather Information Requirement Size
for Air Carrier in 2005.**

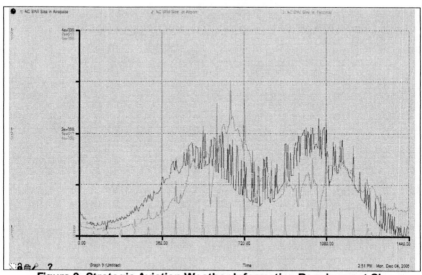

**Figure 9. Strategic Aviation Weather Information Requirement Size
for Air Carrier in 2005.**

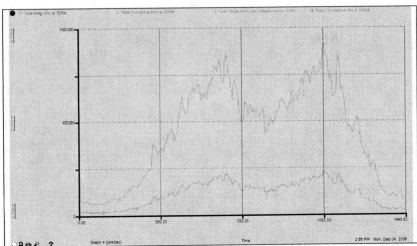

**Figure 10. Aircraft as Sensor in 2005**

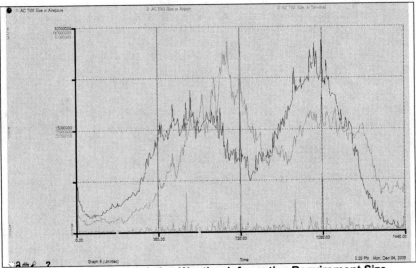

**Figure 11. Tactical Aviation Weather Information Requirement Size
for Air Carrier in 2025**

# An Aggregate Model of Controller Requirements for Enroute Traffic Control

Shin-Lai Tien[1] and Paul Schonfeld[2]

[1] Graduate research assistant, Department of Civil and Environmental Engineering, University of Maryland, College Park, Maryland, 20742. U.S.A. Tel: +1-(301)-405-3160. Email: alextien@umd.edu

[2] Professor, Department of Civil and Environmental Engineering, University of Maryland, College Park, Maryland, 20742. U.S.A. Tel: +1-(301)-405-1954. Email: pschon@umd.edu

## Abstract

This study develops one of the performance modules of the FAA's NAS Strategy Simulator project. An aggregate model that estimates the annual enroute controller requirements for serving the low, mid and high density airspace over the contiguous United States is developed. The current physical boundaries of sectors in each enroute center is neglected while the average sector size and the resulting average sector demand are determined by a theoretical relation, given the peak hour demand of a chunk of airspace and other parameters of airspace characteristics. A statistical relation is then estimated to predict the number of controller positions based on the number of sectors. An iterative procedure connects all components in the model and, after converging, determines average sector size and total number of controllers. The model can provide an estimate of future capacity for evaluating enroute performance and can also be used to examine the current approach to determine annual requirements for controllers.

## Introduction and Model Framework

Generally a system's performance is jointly determined by the demand and the capacity, which is itself determined by available resources. For airspace, the enroute capacity is mainly restricted by the workload of available controllers serving enroute sector traffic and by weather conditions. Uncontrollable weather conditions also result in increased controller workload. As controller labor costs have increased from $82.98 per flight in FY1998 to $137.81 per flight in FY2006 (FAA 2006), it is critical to understand future controller requirements based on demand forecast.

This study aims to estimate the annual controller requirements over the contiguous United States at an aggregate level and only for sector traffic control, excluding controllers in supervisory roles, training roles, and other traffic management positions. The conceptual basis of the model is to neglect the precise actual sector boundaries and to determine the average sector size and the resulting average sector capacity based on peak hour demand, controller capabilities and other system characteristics. The annual controllers needed are then calculated based on the peak hour requirement and other shift and scheduling concerns.

The model is designed to include several components that can be separately considered and improved in future studies, including:

- A method that categorizes airspace based on traffic density.
- A set of staffing standards in units of flights/sector*hour that define the average sector capacity.
- A model that estimates the average open positions (or controllers staffed) per sector.
- An equation derived for calculating average sector size that equalizes sector demand and sector capacity.

After estimating each model for each airspace category, an algorithm connects all the components and iteratively computes the average sector size and the average controllers per sector until the convergence criterion is reached. Figure 1 shows the relation among these components for each airspace category. Each component of the model is described and analyzed in the next section. Estimates of the anticipated relations among variables, and the inputs of the iterative algorithm are described, as shown in Figure 1.

## Model Estimation

### Categorization of Airspace

Ideally, for categorization of airspace, the physical boundaries between regions controlled by FAA facilities are first removed, and the percentages of high, mid, and low density airspace in the NAS are estimated, assuming similar demand patterns over time and homogeneity within each category. However, such an approach is difficult to readily implement, given the availablility of data, which are collected based on existing FAA center and sector boundaries. Until a model that simulates the air traffic movements and generates reliable outputs is validated and calibrated, the data from the 20 ARTCCs over the contiguous United States are used to categorize airspace based on their controlled areas. (Those areas controlled by TRACONs and ATC Towers are not included.)

The k-means clustering method (Johnson and Wichern 2002) is used to group 20 centers based on their similarities in annual operations, delay levels, and areas into three categories summarized in Table 1 and Figure 2. Group 1 covers 17.01% of the airspace, Group 2 32.74%, and Group 3 50.24%.

### Relation between Open Positions and Active Sectors

Sectors and their associated positions (i.e., radar, radar associate, handoff, and flight data) are designed to handle daily traffic. During off-peak periods, the sectors with less traffic are combined with adjacent sectors, and thus the resulting areas served by controllers after combination are enlarged. Several positions associated with low-traffic sectors will also be closed or combined with others and handled by one qualified controller.

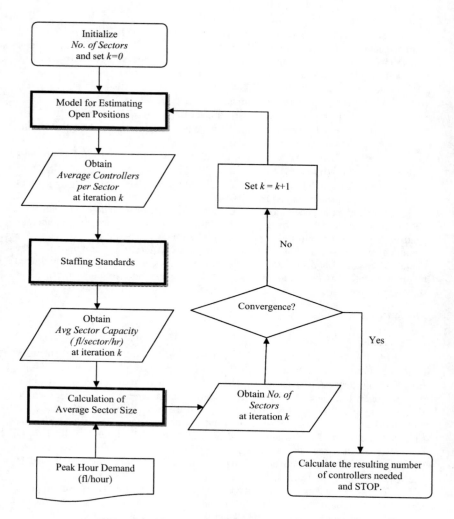

**Figure. 1  Algorithmic Framework of this Study**

The number of hourly open positions actually implies the number of controllers needed in that hour. A statistical relation to forecast the number of open positions based on the number of active sectors should therefore be developed. This relation will be used to forecast the number of controllers needed if the number of active sectors is provided. The premise here is that for each airspace category, given the operational characteristics of current sector design, decisions about expanding the number of open positions can be predicted through a function of the number of active sectors. The individual differences among sectors and the work characteristics of controller positions are overlooked. Note that other considerations could also influence decisions about opening or closing positions and the regression analysis is not the only tool to achieve the above goal. A more complex analysis might be later developed to replace this estimation model.

### Table 1. The Clustering Results of Three Groups of Enroute Centers

| Group (i) | Facility ID | Location | 2005 Operations* | 2005 Avg. Minutes per Delay* | Area** (miles²) |
|---|---|---|---|---|---|
| 1 | ZAU | Chicago, IL | 2,898,206 | 32.52 | 108,064.48 |
| | ZDC | Washington, D.C. | 3,079,365 | 39.21 | 172,552.86 |
| | ZID | Indianapolis, IN | 2,884,052 | 28.99 | 95,353.60 |
| | ZNY | New York, NY | 3,074,207 | 39.32 | 96,476.67 |
| | ZOB | Cleveland, OH | 3,019,900 | 31.67 | 96,183.44 |
| | ZTL | Atlanta, GA | 3,235,785 | 28.32 | 124,526.61 |
| 2 | ZBW | Boston, MA | 1,868,897 | 48.99 | 156,547.69 |
| | ZDV | Denver, CO | 1,836,408 | 39.41 | 266,847.71 |
| | ZFW | Fort Worth, TX | 2,133,793 | 44.96 | 170,819.78 |
| | ZJX | Jacksonville, FL | 2,556,828 | 50.34 | 181,986.54 |
| | ZKC | Kansas City, MO | 2,083,057 | 32.36 | 179,510.62 |
| | ZMA | Miami, FL | 2,500,981 | 34.84 | 227,429.70 |
| | ZME | Memphis, TN | 2,307,707 | 45.53 | 150,639.52 |
| 3 | ZAB | Albuquerque, NM | 1,762,385 | 20.89 | 240,266.35 |
| | ZHU | Houston, TX | 2,121,981 | 27.1 | 344,594.44 |
| | ZLA | Los Angeles, CA | 2,288,328 | 22.15 | 215,786.27 |
| | ZLC | Salt Lake City, UT | 1,551,545 | 24.88 | 416,691.03 |
| | ZMP | Minneapolis, MN | 2,141,528 | 27.1 | 386,583.87 |
| | ZOA | Oakland, CA | 1,703,063 | 20.15 | 181,788.77 |
| | ZSE | Seattle, WA | 1,305,905 | 19.17 | 261,168.99 |

\* Data are from OPSNET.
\*\* Calculated from ETMS boundary point data; Oceanic sectors of ZNY and ZMA are excluded.

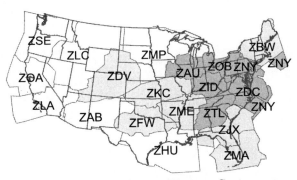

**Figure 2.  Three Groups of Enroute Centers**

By utilizing data from the CRU-Support/ART controller task history database (FAA 2006), we estimate how the number of open positions is expanded if the number of active sectors increases. Two dates, namely 11/03/2005 and 10/28/2005, were selected by the FAA ATO-P Office and CNA Corporation as the representative busiest and least busy days in that season (Gulding and Bonn 2006). The controller tasks that are directly related to handling sectors are identified and selected, and the relation between the combined sectors and combined positions is analyzed. It is expected that the number of open positions will increase non-linearly with the number of active sectors and this property will help the computation to converge. For three groups of ARTCCs, the 15-minute open positions and active sectors are shown in Figures 3, 4, and 5 (in Eastern Standard Time).

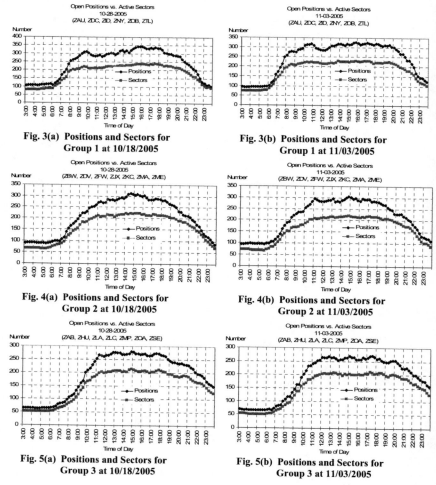

Fig. 3(a)  Positions and Sectors for Group 1 at 10/18/2005

Fig. 3(b)  Positions and Sectors for Group 1 at 11/03/2005

Fig. 4(a)  Positions and Sectors for Group 2 at 10/18/2005

Fig. 4(b)  Positions and Sectors for Group 2 at 11/03/2005

Fig. 5(a)  Positions and Sectors for Group 3 at 10/18/2005

Fig. 5(b)  Positions and Sectors for Group 3 at 11/03/2005

These figures show that the numbers of positions and sectors are quite close at midnight but significantly different during normal operation hours. The estimated relations for the three groups have high statistical significance with R-squared over 0.97 and are summarized in Table 2.

**Table 2. Estimated Functions of Open Positions for Each Group**

| Airspace Category | Function |
|---|---|
| Group 1 | (Open Positions) = 0.7855 * (Active Sectors)$^{1.1023}$ |
| Group 2 | (Open Positions) = 1.2086 * (Active Sectors)$^{1.0140}$ |
| Group 3 | (Open Positions) = 1.2004 * (Active Sectors)$^{1.0107}$ |

### *Derivation of Average (Active) Sector Size*

Equalizing the controllers' workload is one of the considerations in sector design. One of the main features of this model is that the average sector size is determined by considering the peak demand and the average sector capacity. Figure 6 shows how the number of aircraft worked per sector in a 15-minute interval depends on the number of controllers, according to the FAA's recent controller staffing standards. This provides the basis for the estimation of hourly sector capacity in Table 3.

**Number of Controllers by Function and Number of Aircraft Worked**

| Function | Number of Aircraft Worked During 15-Minute Interval | Number of Controllers |
|---|---|---|
| High Altitude Radar Sector | 0 | 0 |
| | 1 - 12 | 1 |
| | 13 - 17 | 2 |
| | 18 - 29 | 3 |
| | 30+ | 4 |
| Low/Both Altitude Radar Sector | 0 | 0 |
| | 1 - 10 | 1 |
| | 11 - 14 | 2 |
| | 15 - 24 | 3 |
| | 25+ | 4 |
| A-Side | 0 | 0 |
| | 1 - 73 | 1 |
| | 74+ | 2 |

NOTE: For application, count aircraft worked for radar sector controller positions during current 15-minute interval. Count aircraft worked for A-side positions at +30 minutes from current 15-minute interval.

**Figure. 6 Effects of Number of Controllers on Sector Capacities (FAA, 1997)**

**Table 3. Average Controllers per Sector vs. Hourly Aircraft Handled**

| | | No. of Aircraft Worked during 1-Hour Interval (flights/sector*hr) |
|---|---|---|
| No. of Controller per Sector per Hour* | 1 | 44 |
| | 2 | 62 |
| | 3 | 106 |
| | 4 | 130 |

*Use interpolation if average controllers per sector are not an integer.

The average sector size can be represented as a function of demand, sector capacity, airspace area, and the average flight distance traveled (Luo and Schonfeld 2003). The derivation of this theoretical relation starts by assuming that the general relation between an area and the distance crossing that area can be represented as:

$$\text{avg distance for crossing through an area} = \text{shape factor} \star \sqrt{\text{area}} \qquad (1)$$

Thus, for airspace category (i), the average flight distance crossing a sector can be approximated by:

$$\text{avg flight distance crossing a sector (i)} = \text{shape factor} \cdot \sqrt{\text{avg sector size (i)}} \qquad (2)$$

Define the average sector passages per flight:

$$\begin{aligned}\text{avg sectors overflown per flight (i)} &= \frac{\text{avg flight distance crossing airspace (i)}}{\text{avg flight distance crossing a sector (i)}} \\[2mm] &= \frac{\text{avg flight distance crossing airspace (i)}}{\text{shape factor} \cdot \sqrt{\text{avg sector size (i)}}}\end{aligned} \qquad (3)$$

Assume that the demand of sector overflights reaches the sector's peak hour capacity for them and the sector size is maximized:

$$\begin{aligned}\text{avg sector capacity (i)} &= \text{avg sector demand (i)} \\[2mm] &= \frac{\text{peak hour demand (i)} \cdot \text{avg sectors overflown per flight (i)}}{\text{avg number of sectors (i)}} \\[2mm] &= \frac{\text{peak hour demand (i)} \cdot \text{avg sectors overflown per flight (i)}}{\left( \dfrac{\text{airspace area (i)} \cdot \text{no. of layers}}{\text{avg sector size (i)}} \right)}\end{aligned} \qquad (4)$$

Rearranging the terms in Eq. (4), we derive the equation of average sector size:

$$\text{avg sector size (i)} = \left( \frac{\text{avg sector capacity} \cdot \text{airspace area (i)} \cdot \text{no. of layers} \cdot \text{shape factor}}{\text{peak hour demand (i)} \cdot \text{avg flight distance within airspace (i)}} \right)^2 \qquad (5)$$

**Model Application**

To demonstrate the practicability of the model, we analyze the difference of airspace structure among 20 enroute centers over contiguous United States and group them based on their similarities in annual operations, delay levels, and other characteristics, e.g. the number of covered airports. Given the assumptions of similar demand pattern over time and homogeneity within each group, the function of open position and other related parameters for each group of enroute centers is statistically estimated. The model is then applied to obtain the peak hour requirement of controllers for each group. The number of annual controllers needed is then calculated based on the peak hour requirement and other shift and scheduling concerns.

Several exogenous parameters must therefore be determined before implementing the model. These are listed below.

- **Avg. flight distance (i)**: Average distance traveled by flights in airspace (i) (assumed; user adjustable)
- **Peak hour demand (i)**: The number of peak hour flights in airspace category (i) (assumed; user adjustable)

- **Shape factor**: This factor is explained in Equation (1) and used to describe the relation between average crossing distance and a given area.
- **Number of layers (i)**: Average layers classified for airspace (i) during peak hours.
- **Shift factor**: This factor is used to convert hourly controllers needed to daily and accounts for daily shift requirement and needs to be further explored. In this study, this value is assumed and user adjustable.
- **Annual staffing factor**: This factor accounts for 7-day facility operation and off-position activities of controllers; 1.7626 is the current value employed by FAA.

Particularly, the shape factor in Equation (1) is used to approximate the average distance needed to pass though an area. To provide with a reasonable input for this parameter, one-day ETMS flight tracks (i.e. October 28, 2005) are analyzed to determine the average shape factor for each sector by calculating the average flight distance divided by square root of sector area. Table 4 summarizes the sector average shape factors for three groups, which are very close to 0.9. These numbers will be applied later to our numerical example.

**Table 4.  Sector Average Shape Factor for Three Airspace Categories**

|                              | Group 1 | Group 2 | Group 3 |
|------------------------------|---------|---------|---------|
| Sector Average Shape Factor  | 0.911   | 0.899   | 0.910   |

The number of altitude layers in any given sector is also a parameter to be estimated. The ranges of flight levels of sectors in each enroute center are analyzed and the criteria for categorizing sector altitude (shown in Table 5) are defined. The current sector design is used for calculating the average layers of sectors in the airspace. A summary categorizing sector altitude for each group of enroute centers and the weights of each altitude category is provided in Table 6. The average number of layers for each group is then calculated: 2.677 for Group 1; 2.611 for Group 2; and 2.528 for Group 3.

With descriptive analyses and reasonable assumptions, the parameters used for implementing the model for three groups of ARTCCs are listed in Table 7.

**Table 5.  Definitions of Sector Altitude in this Study**

|                       | Starting Altitude (the lowest flight level) | Ending Altitude (the highest flight level) |
|-----------------------|---------------------------------------------|--------------------------------------------|
| High Altitude Sector  | > FL180                                     | <= FL999                                   |
| Low Altitude Sector   | >= FL000                                    | <= FL180                                   |
| Mid Altitude Sector   | Between FL000 And FL180                     | Between FL181 And FL999                    |
| All Altitude Sector   | <FL120                                      | =FL999                                     |

**Table 6. Summary of Sector Altitude for Three Groups of Enroute Centers**

| Altitude Category | | High Altitude | Low Altitude | Mid Altitude | All Altitude | Sum of Sectors |
|---|---|---|---|---|---|---|
| Weights | | 2.5 | 2.5 | 3 | 1 | |
| Group | 1 | 123 (38.32%) | 68 (21.18%) | 126 (39.25%) | 4 (1.25%) | 321 (100.00%) |
| | 2 | 131 (40.43%) | 57 (17.59%) | 120 (37.04%) | 16 (4.94%) | 324 (100.00%) |
| | 3 | 103 (34.56%) | 58 (19.46%) | 107 (35.91%) | 30 (10.07%) | 298 (100.00%) |

**Table 7. Summary of Input Values**

| Parameter | Group 1 (i=1) | Group 2 (i=2) | Group 3 (i=3) |
|---|---|---|---|
| **Peak Hour Demand (i)** | 1800 flights/hr | 1600 flights/hr | 1600 flights/hr |
| **Avg. Flight Distance (i)** | 550 miles | 800 miles | 900 miles |
| Airspace Area (i) | 693,157 mile$^2$ | 1,333,781 mile$^2$ | 2,046,879 mile$^2$ |
| Number of Layers (i) | 2.677 | 2.611 | 2.528 |
| Shape Factor | 0.9 | 0.9 | 0.9 |
| Shift Factor | 3 | 3 | 3 |
| Annual Staffing Factor | 1.7626 | 1.7626 | 1.7626 |

*Computational Results*

The exponential functions provided in Table 2 are used to forecast the open positions for the required number of sectors. The results obtained from the algorithm are summarized in Table 8.

**Table 8. Computational Results for Each Airspace Category**

| Airspace Category | Peak Hour Demand (fl/hr) | Number of Sectors | Number of Positions | Average Sector Size (mile$^2$) | Avg. Controllers per Sector | Avg. Sector Capacity (fl/hr/sect.) |
|---|---|---|---|---|---|---|
| Group 1 | 1800 | 258.06 | 357.75 | 7,192.13 | 1.39 | 50.26 |
| Group 2 | 1600 | 256.57 | 335.13 | 13,573.94 | 1.31 | 47.58 |
| Group 3 | 1600 | 229.37 | 291.82 | 22,564.44 | 1.27 | 46.44 |

Given the peak hour demand for each group of centers, the annual requirements of enroute controllers are estimated as:

Annual Requirement of Enroute Controllers
= Total Number of Controller Positions in Peak Hour * Shift Factor * Annual Staffing Factor
= ( 357.75 + 335.13 + 291.82 ) * 3 * 1.7626 ≈ 5207

Note that this estimate of annual enroute controller requirements accounts for only enroute sector control since the balance of sector demand and sector capacity is the main consideration. Other roles in traffic management unit in an ARTCC will require further exploration. When used for comparison with historical controller staffing data, the results of this model should be carefully interpreted since many operational issues and culture differences among enroute centers are not considered here.

For the model to become a more realistic tool of policy analysis, the following inputs should be carefully identified before the model is applied:

- Peak hour demand for each group of enroute centers,
- Avg. flight distance in each group of enroute centers, and
- A shift factor that converts hourly controllers needed to daily and accounts for daily shift requirement.

**Concluding Remarks**

In this study, various methods are explored and used to analyze sector boundary data, to derive a sector size formula, and to estimate the relation between open positions and active sectors. This is required to obtain proper model inputs and to demonstrate the applicability and practicability of the proposed model. The aggregate level annual enroute controller requirements generated by this model can provide the FAA and other civil aviation organizations worldwide with a practical tool for estimating staffing requirements in response to future traffic growth. This model can be also extended to estimate controller costs or system performance with changes in technologies or operating procedures.

**References**

Luo, Y. and P. Schonfeld (2003). *Airport Capacity and Airspace Performance Module – Phase II Report*, University of Maryland, College Park.

Gulding, J. and Bonn, J. (2006). *Analysis of Representative Days*, FAA ATO-Office of Strategy.

Federal Aviation Administration (1997). *FAA Staffing Standards Report: ARTCC Radar Sector Staffing Models*.

Johnson, R. and Wichern, D. (2002). *Applied Multivariate Statistical Analysis*, Prentice-Hall.

Federal Aviation Administration ATO-P Office. CRU-Support/ART Database. Retrieved on August 29, 2006.

Federal Aviation Administration (2006). *Submission to the United States Congress Concerning the Agency's Collective Bargaining Proposal to the National Air Traffic Controller Association*.

# Washington Dulles Airfield Development - Preparing for the Future

Gary K. Fuselier[1], Michael Hewitt, P.E.[2] and Joseph S. Grubbs, P.E.[3]

[1]Project Design Manager for Metropolitan Washington Airports Authority, Ronald Reagan Washington National Airport, Washington, DC 20001-4901; PH (703) 417-8189; email: gary.fuselier@mwaa.com

[2]Senior Civil Engineer for Parsons Management Consultants, 45045 Aviation Drive, Suite 300, Dulles, Virginia 20166-7528; PH (703) 572-1106; email: Michael.hewitt@mwaa.com

[3]Senior Aviation Engineer for CH2M HILL, 200 Corporate Center Drive, Suite 150, Moon Township, PA 15108; PH (412) 604-4043; email: jgrubbs@ch2m.com

## Abstract

As the aviation industry continues to change and the airlines strive to return to profitability, airport owners are faced with a variety of infrastructure challenges to provide capacity for future industry growth. The Metropolitan Washington Airports Authority has undertaken an aggressive development program to position the Washington Dulles International Airport (IAD) to meet these challenges. IAD is in the midst of a $3.4 billion development program, named Dulles Development ($D^2$). The focal point of this paper is airfield capacity enhancements and improvements at IAD. The paper presents the steps taken to increase the airfield capacity with the addition of a new runway complex, Runway 1L-19R, and the implementation of taxiway geometric improvements for the A-380 aircraft at the Runway 1R-19L complex. The paper also discusses other major airfield improvements such as the reconstruction of Runway 12-30 and Runway 1C-19C. The paper examines the origins of the aforementioned projects, the implementation and progress of the airfield development portion of the $D^2$ program and the expected future benefits of the program in providing the IAD airfield with the flexibility to adapt to an ever changing aviation industry.

## Airport Overview

IAD is located in Loudoun and Fairfax Counties in northeastern Virginia, approximately 26 miles northwest of Washington, D.C. The Metropolitan Washington Airports Authority (the Authority) owns and operates the facility. IAD is designated as a large hub, primary commercial service, airport in the FAA's National

160

Plan of Integrated Airport Systems (NPIAS). IAD serves as the "growth" airport in the metropolitan Washington D.C. region, since Ronald Reagan Washington National Airport is constrained by federal legislation and surrounding land uses. Baltimore Washington International Airport (BWI) also serves the greater Washington D.C. region but is similarly encumbered by land constraints.

Three Portland cement concrete (PCC) runways serve the airport. These runways are designated 1L-19R, 1R-19L and 12-30. Runways 1L-19R and 1R-19L are each 11,500 feet long by 150 feet wide. Runway 12-30 is 10,500 feet long by 150 feet wide. With the volume of air travel continuing to increase, the Authority continues to add gate space to accommodate this volume. Keeping pace with the increase of terminal capacity, several airfield enhancements have been undertaken, including construction of a third CAT II/III parallel runway (future Runway 1L-19R), taxiway geometric improvements for the A-380 aircraft at the Runway 1R-19L complex, reconstruction of the cross wind runway (Runway 12-30) and the reconstruction of existing Runway 1L-19R (soon to be re-designated as Runway 1C-19C).

**Eyeing Future Capacity – Construction of New CAT III Runway**

As the owner and operator of IAD, the Authority has embarked on a program to expand and improve the airport's facilities to ultimately accommodate nearly 55 million passengers per year. Central to this goal is a project to build a new air carrier runway and associated facilities. The purpose of the new runway is to safely and efficiently increase the capacity of the airfield to accommodate the growing number of aircraft operations and to provide the airport with flexibility for handling these additional operations. The new fourth runway is consistent with the preferred development and master plans for a five runway system at IAD, which includes 2 East-West Parallel Cross wind Runways and 3 North-South Parallel Runways. The addition of the fifth runway will be capacity driven. Its addition is currently unscheduled.

The Authority is seeking to continue to provide mid- and long-term operational capability to accommodate the forecasts of aircraft operations in a flexible, safe and efficient manner. The greatest increase in runway and airfield capacity would be achieved through a design that could 1) provide needed capacity and operational throughput to accommodate the projected levels of aircraft operational demand; 2) accommodate future levels of operational demand without incurring unacceptable levels of aircraft delay; and 3) provide the capability to accommodate triple simultaneous independent operations in north-south flows during instrument meteorological conditions. The fourth runway and associated improvements at IAD will provide significant capacity improvement to the airfield by meeting these goals.

**Fourth Runway Complex**

The new fourth runway, which will be designated Runway 1L-19R, is being constructed on a site west of the Dulles Airport terminal area consisting of old farms, mature and new forest, and open pasture land (see Figure 1). Two streams cross the project site and numerous old abandoned houses and farm structures have been demolished.

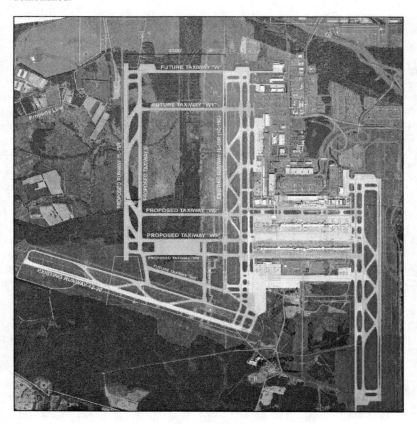

**Figure 1. Fourth Runway Location**

The new PCC runway is 9,400 feet long and 150 feet wide with 35-feet wide bituminous concrete paved shoulders. It is located 4,300 feet west of and parallel to the existing Runway 1L-19R (to be re-designated 1C-19C). The new runway is to be used by a variety of air transport category aircraft ranging in size from commuter or regional aircraft to large aircraft used for transoceanic or other international flights. The runway will be lighted and have new Navigational Aids to ultimately accommodate Category II/III approaches to both runway ends.

All taxiways will be constructed 75 feet wide of PCC and have 35 feet wide bituminous concrete paved shoulders.  The project includes a parallel taxiway, connector taxiways, and cross-field taxiways that connect the new runway to the airport terminal area and the rest of the airfield.  Four, 30 degree high speed exit taxiways (two for northbound arrivals and two for southbound arrivals) are also being constructed. The locations of the high speed exits were modeled to optimize the average runway occupancy times and also to provide a direct connection to the cross field Taxiways W1 and W2. After the initial phase of construction, the parallel taxiway will include four entrance/exit taxiways: Taxiways W2, W3, W4 and T. Two additional cross field taxiways (Taxiways W and W1) are planned from the new north side of the parallel taxiway complex with the possibility of these being constructed with the Center Runway Reconstruction discussed below. Future taxiway improvements are also planned to include taxiways south of Taxiway W4, which will include the construction of Taxiways S and W5. These taxiways will provide further capacity improvement by providing access points to the Runway 1C threshold and a cross field capability to Taxiway Q, Runway 12-30 and the future fifth runway.

In addition to the new runway and taxiways, a new remote deicing pad "B" and deicing facility will be constructed to provide IAD the capability to remotely deice aircraft operating on Runway 1L-19R and Runway 1C-19C, and as such, will provide the airfield with capacity improvements when queuing and deicing aircraft. The remote pad is also planned to be utilized as a hold pad during non-deicing periods and will provide the airfield with capacity improvements when queuing and deicing aircraft. The further flexibility of the airlines to move aircraft from the gates will reduce the potential for aircraft delay when arriving and departing from the terminal area. The new deicing pad was designed with the flexibility to accommodate wide body, narrow body and commuter jet traffic.

**Project Construction**

To prepare the project site for pavements and navigational aids, a significant clearing and demolition effort was implemented (Fuselier and Grubbs 2004). Approximately 600 acres of trees were cleared and numerous structures and utilities were demolished or relocated. Approximately 90 acres of wetlands were filled, and two perennial streams were diverted into large box culverts passing underneath the runway structure, all in accordance with the environmental permits obtained for this project. Grading consisted of approximately 2.5 million cubic yards of earthwork excavation and 675 acres of topsoil stripping.

The new Runway 1L-19R will be paved to conform to the Federal Aviation Administration (FAA) criteria for aircraft design group V with blast pads 400 feet long by 220 feet wide. The pavement section consists of 17 inches of PCC pavement, 6 inches of cement treated base course and 12 inches of cement stabilized subgrade. The separation between the proposed runway and parallel taxiway is 600 feet, as recommended by the FAA for runways with acute angled exit taxiways. Approximately 92 acres of PCC paving will be completed with the first phase of construction.

The new lighting systems will require the installation of over 130 miles of cable and over 2,500 edge and in-pavement airfield lighting fixtures. Navigational aid improvements include new CAT II/III glide slopes and localizers, inner markers and far field monitors for both runway ends and a new touchdown and midfield runway visual range. Additional visual aids include new four box PAPI systems and new ALSF-2 high intensity approach lighting systems for each runway end. Supporting facilities include ATCT control and monitoring equipment, a fiber optic transmission system supplemented by a copper communications cable system between the field NAVAIDs and the new ATCT, and a new electrical Vault 4 facility.

The stormwater management system for the future runway system and de-icing pad will be constructed to take into consideration environmental resources, stormwater regulatory requirements, and long-term airport operational needs. The proposed stormwater treatment system is designed to detain stormwater for treatment through passive biological treatment systems for removal of BOD from fugitive deicing materials and total phosphorus. Low Impact Development (LID) concepts were incorporated into the design to meet Virginia Department of Environmental Quality (VDEQ) water quality requirements. The principles of LID require that runoff be minimized by promoting infiltration and treatment onsite or as near the source as possible. Approximately 4,600 LF of single, double and triple box culverts and more than 15,000 LF of storm pipe will be constructed. Drainage design considerations included numerous crossings of the mapped floodplain and accommodations for future development in newly accessed areas of the airport.

As the new runway facility will be built in an essentially "green" site away from existing airport facilities, a significant perimeter and access roadway plan was developed. Included will be a network of paved and gravel roadways and access points to the fourth runway complex. A new two-lane vehicular tunnel under Taxiway W2 will be constructed to eliminate conflicts between ground support equipment and aircraft operations and to provide free flowing access to the deicing facility, where deicing materials and fugitive glycol will be stored. A new perimeter fence and security gate system will be constructed on the west side of the airport, securing the facility and the new land on which the runway will be constructed.

Ancillary facilities include a new ARFF station on the northern side of the project site to provide primary response capability to the Fourth Runway; and a new support facility which will be used to store equipment, materials and house staff necessary to maintain and operate the new runway complex.

When completed in the fall of 2008, the Fourth Runway complex at IAD will provide the Authority with a significant improvement in the capacity and functionality of the airfield. The investment in designing, constructing and operating the Fourth Runway will provide the Washington D.C. region and the greater national aviation network with needed flexibility and capacity to operate with fewer delays and to account for future growth.

**Accessibility for NLA - A380 Airfield Study and Implementation**

As the primary international gateway to our Nation's Capital, IAD is anticipating the addition of the Airbus A380 aircraft to its airfield traffic mix by at least three air carriers. In preparation for the arrival of the A380, a study (Crawford, Murphy & Tilly, Inc et al. 2005) was conducted in 2005 to quantify the airfield modifications necessary to support this aircraft. The study evaluated the existing runway and taxiway geometry, pavement markings, lighting, signage, NAVAIDs and the runway pavement edge strength for compliance with Design Group (DG) VI and other FAA criteria for the A380. A separate study focused on evaluating modifications to the parking positions, gates and concourses.

The Airfield Study (Crawford, Murphy & Tilly, Inc et al. 2005) evaluated each of the three runway complexes for full compliance with available FAA Design Group VI criteria (as modified by Engineering Briefs 63 and 65). Given the published performance criteria of this aircraft, dedicated travel paths for each runway complex on the airfield were established for the A380 (see Figure 2). For these travel paths, the following items were evaluated:

1. Pavement geometry, pavement markings and taxi routes analysis included the size and shape of the available paved areas and the ability of the A380 to transit these areas.
2. Airfield lights, airfield signage and NAVAIDs analysis examined the existing location of the runway and taxiway edge, centerline and approach lights.
3. Runway edge pavement strength evaluation included Non-Destructive Testing and strength analysis of the runway shoulder pavement for comparison to the FAA required strengths per Engineering Brief 65.

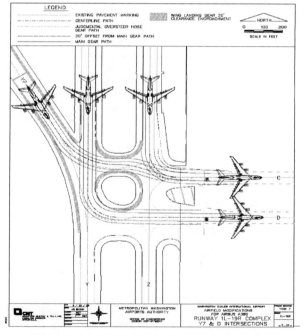

**Figure 2. A380 Pavement Geometry Evaluation**

The study identified a variety of necessary modifications to facilitate the A380. These modifications were quantified for programming future construction improvements and associated costs for each of the three existing runway complexes in supporting the Airbus A380 operations. Given all of the information, Runway 1R-19L was selected as the primary runway to support the A380.

Implementation of several aspects of the study has commenced. One recently completed construction project constructed the exit taxiway (Taxiway K7) geometric enhancements to support arrivals of the A380 on Runway 19L. A second and similar project on Taxiway K2 is slated for construction this year to support arrivals of the A380 on Runway 1R.

## Reconstruction of Existing Runways

In order to handle the larger heavier modern day aircraft, it became necessary to reconstruct the PCC pavement of Runway 12-30 and Runway 1L-19R (soon to be re-designated 1C-19C). With the increased traffic at IAD in recent years, these original airport pavements, constructed in the early 1960's, have been deteriorating at an accelerated rate.

## Reconstruction of Runway 12-30

The project consisted of reconstructing the full width of 10,000 feet of the total 10,500 feet PCC pavement. The reconstruction included demolition of the existing 15-inch PCC pavement and replacement with an 18-inch PCC pavement section; isolated PCC slab repairs on the adjacent taxiways; pavement grooving; and shoulder work to allow for the pavement construction. The shoulder work also permitted the installation of pavement underdrains. With the replacement of the pavement section, the runway centerline and touchdown zone lighting were also replaced. These lighting systems were drained using the pavement underdrain system.

Runway 12-30 (see Figure 3) is the only cross wind runway at IAD. The larger, heavier aircraft have been using Runway 12-30 with increased frequency. Additionally, there are certain times of the year where the wind conditions dictate that smaller aircraft (Group III or smaller) use Runway 12-30. The critical nature of Runway 12-30 to airfield operations underscored the importance placed on reopening the runway as quickly as possible.

Given the aforementioned wind condition, the runway reconstruction had to be completed within a four month window (June to September). Extensive modeling was conducted to quantify the capacity impacts to the airport of closing the runway. Using a Stakeholder Buy-in Process[2], several construction phasing options were evaluated to complete the construction within this operational window. Ultimately, the design team recommended and received Stakeholder buy-in to close the runway for its entire length during reconstruction. The design was prepared to facilitate an aggressive construction schedule, setting the contract time at 134 calendar days for the reconstruction. Additionally, incentive language for early completion was provided in the contract documents to entice the contractor to better the 134 calendar days. The costs incurred by the airport as a result of closing the runway were calculated to be in excess of $100,000 per day. The incentive was $100,000 per day for early completion of work up to a maximum of $1,500,000.

**Figure 3. Runway 12-30 PCC Construction**

Construction commenced in the summer of 2004 and had to overcome several obstacles, most notably a regional cement shortage. Between the times the project was advertised for construction and the actual start of construction, the Mid-Atlantic region of the United States was facing a cement shortage. At the time, approximately 23% of the cement used in the United States was imported. A large portion of this imported cement was being diverted to other nations, causing a cement shortage on the east coast. This shortage directly affected the Runway 12-30 Reconstruction Project. Since the demand for cement greatly exceeded the supply, cement suppliers were only willing to guarantee cement quantities for a relatively short period of time. By the time the construction contracts were awarded, the guarantee period had expired. The cement slated for this project went to other regional projects, leaving the Runway 12-30 project without any firm commitments for either the source or quantity of cement.

Since the contractor, the Engineer and the Authority were members of the American Concrete Pavement Association (ACPA), ACPA became instrumental in assisting the contractor in obtaining the majority of the cement needed for the project. Realizing this was a very high profile project, on a very tight production schedule, the ACPA immediately contacted the cement producers in the region to develop a feasible solution to the cement shortage.

The original mix design was changed to utilize Ground Granulated Blast Furnace Slag (GGBFS). This change eased some of the cement needs of the project. Following this change, the majority of the demand for the remaining cement was met by two of the cement producers contacted by ACPA.

However, even with this assistance, there was still an insufficient amount of cement available to complete the project as originally designed. The original design for this pavement section of the runway consisted of PCC pavement on a cement treated base course (CTBC) with a soil cement stabilized subgrade, the standard components of airfield pavement at IAD. By remaining intimately involved during the construction, the design team was able to offer a solution to the cement shortage by replacing the CTBC with a bituminous base course. The cement quantity reduction was more than enough to complete the remaining courses as originally designed. With a quick turnaround on the pavement structure analysis and the preparation of the necessary specifications, the Authority approved the change and approximately 70% of Runway 12-30 was constructed with a bituminous base course (see Figure 4).

**Figure 4. Bituminous Base Course for Runway 12-30**
Note: Because of the cement shortage, the construction team switched from cement treated to bituminous base course (125,000 square yards), saving approx. 4,000 tons of cement

Overcoming this obstacle was critical in re-opening the runway. The runway was successfully placed back into full operation in 121 calendar days, well within the operational window of four months. By re-opening the runway in this time frame, the contractor achieved thirteen of the fifteen incentive days for a total bonus of $1.3 million.

## Reconstruction and Re-designation of Runway 1C-19C

The project includes the design of the future Center Runway and future Taxiways V, W and W1. The future center runway is the existing Runway 1L-19R and will be re-designated Runway 1C-19C once the fourth runway opens in the fall of 2008. Taxiways W and W1 will connect the new Fourth Runway parallel taxiway, Taxiway U, to Runway 1C-19C. Taxiways W and W1 will each be approximately 3,400 feet in length. Taxiway V will extend from the eastern edge of Runway 1C-19C to the western edge of Taxiway Z. The project also includes the reconstruction of High Speed Exit Taxiways Y3, Y4, Y5, Y6 and needed repairs on the remaining exit taxiways within the Runway Safety Area.

The reconstruction effort includes new PCC pavement for the runway (threshold to threshold) and taxiways. The runway and associated exit taxiways reconstruction work includes pavement demolition, subgrade preparation, replacement of a portion of the existing bituminous concrete shoulder pavement necessary to facilitate the PCC pavement construction, grade tie-ins and placement of pavement underdrain system. The runway reconstruction also includes replacement of the pavement marking, bituminous concrete blast pads and the in-pavement lighting features (pavement condition sensors as well as centerline, leadoff and touch down zone lighting). The project also includes a confirmation study of a cross field roadway tunnel connecting the terminal midfield area to the area between Runway 1C-19C and Runway 1L-19R. Pending the results of the study, the segment of this tunnel under Runway 1C-19C may be constructed under this project.

The construction of new Taxiways V, W and W1 includes a substantial amount of embankment placement, two-lane roadway tunnels and large barrel box culverts beneath Taxiways W and W1, placement of a PCC pavement section with bituminous concrete shoulders, airfield lighting and signage, other stormwater culverts, and erosion and sediment control.

Packaged into this project will be the additional connection taxiways to the new Fourth Runway. These taxiways, Taxiways W2, W3, and W4, will also be constructed of PCC pavement with a bituminous concrete shoulder and will include new airfield lighting and signage, as well as substantial modifications to the existing airfield signage and electrical circuit routing. The construction of this project is slated to start in late 2008 or early 2009, after the opening of the new Fourth Runway.

## Conclusion

The Metropolitan Washington Airports Authority continues the aggressive $3.4 billion $D^2$ development program to position the IAD to meet the infrastructure challenges to provide capacity for future aviation industry growth in our Nation's Capital. The new Fourth Runway will safely and efficiently increase the capacity of the airfield to accommodate the growing number of aircraft operations and to provide the airport with flexibility for handling these additional operations. The thicker pavement sections of reconstructed Runway 12-30 and Runway 1C-19C will be more serviceable for the increased traffic and heavier aircraft fleet mix. The implementation of the airfield modifications for the Airbus A380 will accommodate several international air carriers in utilizing Airbus A380 service from IAD to destinations around the globe.

**Acknowledgements**

A special thanks to the following for their contributions in the preparation of this paper:

David Stader, P.E. Robert Aycock and Terri Shipley of CH2M HILL
Stan Herrin, P.E. of Crawford, Murphy & Tilly, Inc.
David Field, P.E. of Hatch Mott MacDonald, LLC

**References**

Crawford, Murphy & Tilly, Inc., Roy D. McQueen & Associates, Ltd. and The Burns Group (2005). *Washington Dulles International Airport Airfield Modifications for Airbus A380,* Final Report.

Fuselier, G.K. and J.S. Grubbs (2004). "Building Flexibility Into Airfield Construction Phasing Plans," published in the *American Society of Civil Engineers 28th International Air Transport Conference proceedings*, July 2004.

# Airport Relocation Program for the Panama City-Bay County International Airport

Darin Larson, P.E.[1]

[1] Division Manager, Aviation Services, PBS&J, 1514 Broadway, Suite 203, Fort Myers, Florida 33901 ; PH (239)-334-7275, FAX (239)-334-7277 ; email: drlarson@pbsj.com

## Abstract

The relocation of the Panama City-Bay County International Airport to a new site about 25 miles northwest of Panama City, Florida, exhibits demanding site conditions and poses certain design and permitting challenges. In order to make the project compatible with all aspects of the community, the Design Team envisioned and proposed unique solutions to the new Green Field airport's permitting and mitigation processes. The paper describes a new airport that is planned and permitted around an ultimate 4,000-acre development footprint including two parallel commercial service runways and a crosswind general aviation runway. This project involves using a groundbreaking ecosystem team permitting (ETP) process in which state permits and approvals pertaining to wetlands, storm water management, water supply, wastewater, air quality, threatened and endangered species, and air permits are addressed in a holistic manner. The ETP approach involves a coordinated process whereby the project design and conditions of approval are presented, discussed, and negotiated in an open public forum, resulting in net ecosystem benefits and allowing greater flexibility in project design, issue resolution, and permit issuance.

## Introduction

The current airport is located between North Bay and sprawling suburban development just outside of Panama City. Opportunities for expansion at the current facility are limited. The Airport's primary runway (14-32) at 6,304 feet in length is one of the shortest runways used by commercial airlines in Florida. The Runway Safety Area (RSA) for runway 14-32 does not meet current FAA standards. The Panama City-Bay County International Airport (PFN) site also is so closely bordered by residential development that it is generally considered non-compatible with airport activities.

For these reasons, airport management is taking action to address the Airport's shortcomings. During the 1990's, an Environmental Assessment was initiated to consider alternatives to provide an 8,000 foot runway at PFN. This study recommended an extension of Runway 14-32 to the northwest into Goose Bayou to provide additional length for the Airport's primary runway. However, the proposed runway extension would have environmental impacts to state-protected Class II waters. Due to the opposition for this concept, the Environmental Assessment was deferred in 1998.

In 1999, with support from the Federal Aviation Administration (FAA) and the Florida Department of Transportation (FDOT), the Panama City-Bay County International Airport Authority initiated a new study to establish the feasibility of either expanding or relocating the airport facilities. The Airport Feasibility Report, released in July 2000, documents the findings of that study, and recommended the relocation of PFN rather than expansion of the current site.

## Site Selection

Identification of potential sites for a new airport in Bay County proved to be challenging. Most of Bay County was immediately off-limits due to military airspace and hurricane storm-surge criteria and the search was quickly narrowed to two areas. Area One was in the northwest portion of the county above West Bay, which is connected to North Bay and is part of the larger St. Andrews Bay ecosystem. It is bounded by Eglin Air Force Base's Restricted Airspace to the west and Tyndall Air Force Base's Military Operations Airspace (MOA) to the east. Area Two is in the east part of the county to the northeast of Tyndall Air Force Base, north of State Road 22, between Tyndall's MOA and Class D airspace. When Area Two was eliminated, site identification and evaluation focused on Area One.

Three development sites were analyzed within Area One and a variety of aviation and non-aviation issues were considered. Aviation issues included airspace, wind coverage, terminal approaches, obstructions, and wildlife attractants. Non-aviation issues included, but were not limited to, environmental impacts, development costs, land use and ground access (Panama City- Bay County International Airport 2006a, 2007b). In order to identify and consider potential environmental impacts associated with the potential sites, there was significant coordination with The Florida Department of Environmental Protection (FDEP), The United States Army Corps of Engineers (USACE), and The United States Fish and Wildlife Service (USFWS). The chosen site (see Figure 1 on next page) was narrowed to a 4,000 acre footprint that could accommodate the proposed development for the new airport over at least a fifty year planning horizon, including plans for parallel commercial service runways at 10,000 and 12,000 foot ultimate lengths and a 5,000 foot general aviation crosswind runway.

Typically, one of the most difficult issues associated with any major airport development project, and especially in the case of a new airport, is acquisition of the necessary property. In this case, the selected site was situated in the middle of over 75,000 acres of land owned by the St. Joe Company. In a classic example of a public/private partnership, St. Joe agreed to work with the Panama City – Bay County International Airport and Industrial District toward the development of the new airport. The partnership began with an innovative Sector planning process which established future land use overlay for the long range growth of north-central Bay County including provisions for the new airport, a 3,700-acre Regional Employment Center, an Airport Business Center and tracts for residential land uses of various allowable densities. By planning the development around the proposed airport site, the team was able to create buffers of commercial land use around the airport thereby protecting the ultimate planned development of the airport from encroachment by incompatible land uses, such as residential.

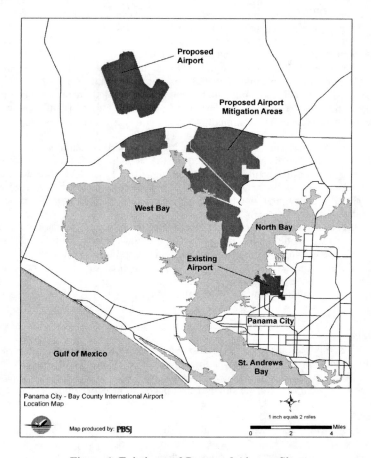

**Figure 1. Existing and Proposed Airport Sites**

One additional unique aspect of the plan is that more than half of the total land is designated as preservation, including 62.5 square miles of bay shoreline and watershed. With respect to the site for the new airport, St. Joe has agreed to donate the 4,000 acres of land for the proposed new airport along with the provision of over 9,600 acres of land for the mitigation necessary to address the environmental impacts from the airport development (http://www.joe.com). Some of these environmental impacts and the associated plan for mitigation are addressed in the following paragraphs.

**Wetland Impacts**

Of the 4,000 acre footprint in the selected site, more than half consisted of jurisdictional wetlands. In addition, over 7,279 linear feet of streams would be impacted by the project. In order to address these wetland impacts, the project team completed a mitigation planning and design plan in the West Bay Preservation Area (WBPA). This area was chosen in order to protect the water quality and resources of the bay and to provide large, contiguous, and essential wildlife corridors.

The project will result in a significant amount of impact to terrestrial and aquatic habitats in wetland and upland pine plantations that exist on the proposed site. However, the loss of wildlife habitat is mitigated through the proposed enhancements within the wetland mitigation area, including implementation of the draft Wildlife Management Program (where practical) and through preservation of native habitats as proposed in the Sector Plan.

The FAA initiated Essential Fish Habitat (EFH) consultation with the National Marine Fisheries Service (NMFS) in a letter dated June 8, 2005. The FAA determined that the proposed airport would have minimal impact on Essential Fish Habitat and that any secondary impacts that may occur would be adequately mitigated. NMFS reviewed the EFH Assessment and determined that they had no further EFH conservation recommendations to offer.

In comparing the proposed new airport to the existing, it is anticipated that the impacts due to potential spills or leaking storage tanks may be less for the new site because the design for the site takes into account precautionary measures to minimize impacts from accidental fuel discharge. These measures include: centralized fueling facilities and above-ground storage tanks, and piping systems with secondary containment. This allows for early detection of leaks or spills and easier repair of the system.

**Floodplains**

The proposed new airport will also result in significant floodplain encroachment. The floodplain impacts would occur in flood zone A within the floodplains of Kelly Branch, Bear Bay, and Morrell Branch for the proposed airport site in the amount of 207.1 acres as compared to the floodplain impacts from redevelopment of the existing airport site in flood zone A, flood zone AE, and flood zone VE and which would total 140 acres.

The FAA determined that due to safety, operational, and engineering siting requirements, there is no practicable alternative to floodplain encroachment at the proposed new airport site. According to the Conceptual Storm Water Master Plan prepared by the airport consultant and as defined in the ETP process, floodplain compensation is not required. Other than construction standards that require protection of buildings from flooding, Bay County does not have a program specifically regulating activities in Federal Emergency Management Act (FEMA)-regulated floodplains. Nonetheless, the airport has committed to design the storm water management system to match pre-project discharge rates and outfall to existing discharge points.

With implementation of the mitigation described above as well as the proposed mitigation for wetland impacts, adverse impacts on natural and beneficial floodplain values have been appropriately offset and will not result in a considerable probability of the loss of human life or the likely future damage associated with the encroachment that could be substantial in cost or extent, including interruption of service or loss of vital transportation facility, or notable impact on natural and beneficial floodplain values.

## Plant and Wildlife Impacts

In comparing the impacts to wildlife that would result form redevelopment of the existing airport site to those at the proposed new airport, the FAA determined that depending on the specific type of development ultimately realized, redevelopment of the existing airport could result in potential impacts to federally-listed species (Eastern indigo snake, several sea turtle species, the manatee, and gulf sturgeon) and state-listed species (gopher tortoise, Florida pine snake, gopher frog, various wading bird species, and the spoon-leaved sundew). One listed plant species, the spoon-leaved sundew, was identified at the airport study area, but it is not located within the area of potential impact for the extension of Runway 14-32. This species could be impacted if the existing aiport were redeveloped.

A Biological Opinion (BO) was issued by the United States Fish and Wildlife Service (USFWS) to address the impacts associated with the proposed new airport. In addition, a BO for the Regional General Permit (RGP) addresses potential impacts to listed species for the subset of the Sector Plan area included in the RGP. As development occurs within the Sector Plan, each development project will be subject to review and approval by various regulatory agencies and avoidance and minimization of impacts to listed species will be evaluated at that time.

Several state-listed plant species will be lost at the proposed airport site by dredging and filling activities. However, these species are relatively abundant on the proposed mitigation parcels and enhancement activities such as controlled burns and thinning of pine canopy should allow these species to flourish.

The FAA initiated consultation with the U.S. Fish and Wildlife Service (USFWS) and the National Marine Fisheries Service (NMFS). The FAA prepared and submitted a Biological Assessment (BA) showing the potential impacts to federally listed species for the proposed airport site. The USFWS reviewed this document and concurred with the FAA determination regarding affects to listed species, including a determination of "may affect, likely to adversely affect" for the flatwoods salamander breeding habitat. The USFWS has subsequently completed a Biological Opinion (BO) for the flatwoods salamander and concluded that the relocation of the airport to the new site would not jeopardize the continued existence of this species. The USFWS determined that the proposed enhancements within the mitigation parcels could actually benefit the salamander.

## Permitting and Mitigation

The acquisition of state and local environmental permits required for this Airport relocation was coordinated in an innovative Ecosystem Team Permitting (ETP) process. This is a process unique to the state of Florida, and it replaces individual permit applications, allowing projects and associated ecosystem impacts and benefits to be viewed and considered as a whole, maximizing benefits and providing a uniform, holistic plan for development and preservation.

ETP, formally authorized under Section 403.0752 Florida Statutes (F.S.), is a process by which the Florida Department of Environmental Protection (FDEP) takes a lead role in coordinating the issuance of all required state environmental permits under a one review.  In addition, federal agencies are included in the process by simultaneously coordinating with the state permit review. The ETP process is initiated through a legal agreement between the Florida Department of Environmental Regulation (FDEP), the owner and other permitting agencies; and is conducted through a series of roundtable meetings with the participating permitting agencies and other stakeholders.

In early 2002, a draft ETP agreement was prepared and submitted to the FDEP. After revising the agreement, a memorandum of agreement for the ETP process was signed in early November 2002.  The Ecosystem Management Agreement itself was issued in early December 2006.

This agreement included a restoration plan for the 9,600 acres of mitigation property and includes the removal of what is now an industrial silviculture plantation and restoration of this area to a wet longleaf pine savanna and flatwoods system.  The restoration will ensure that the land returns to a historic (of the late 1940s and early 1950s) natural ecosystem.  It also will promote preservation of large, contiguous tracts for wildlife corridors and watersheds.  The plan will include enhancements of wetlands, streams and watersheds and long-term preservation of a significant portion of the West Bay mangrove coastline.

An ecological functional assessment of the mitigation plan has demonstrated a significant net gain in wetland function relative to the proposed wetland impacts. After independent review, the FAA supports the USACE's determination that the conceptual mitigation strategy for the Airport Sponsor's Proposed Project would provide sufficient wetland functional lift to offset the proposed wetland functional loss expected from the direct impact to wetlands. Because impacts from the Airport will occur incrementally, (please see Table 1) so will the mitigation plan.  Parcels have been divided into 42 management units of 200 to 300 acres each.

## Table 1 – Incremental Mitigation Plan

| Construction Phase | Management Units | Functional Lift | Total Acreage[1] |
|---|---|---|---|
| 0-10 years (I) | 1B,1C,1F,1G,1H, 2B, 2C,2D,2E,2F,[2] 2M[3], 2P[4],2Q,2R,2S,2V,2Y | 653 | 4039.2 |
| 11-20 years (II) | 2X,3A,3B,3C,3E,3G | 212 | 1334.3 |
| 21-30 years (III) | 1E,2H,2I,2J,2K,2M[3] 2N,2O,2P[4],2T,3D,3F,3H | 414 | 2183.5 |
| 31-40 years (IV) | 2A,2F[2], 2L,2U | 148 | 712.7 |
| 41-50 years (V) | 1A,1D,1I,2G,2W | 296 | 1339.1 |
| Total | NA | 1723 | 9608.8 |
| 1. Total Acreage units includes uplands and wetlands with no functional lift | | | |
| 2. 324.4 Acres of Management Unit 2F are applied to Phase I; 102.9 acres are applied to phase IV. | | | |
| 3. 167.1 acres of Management Unit 2F are applied to Phase I; 3.9 acres are applied to Phase III. | | | |
| 4. 211 acres of Management Unit 2P are applied to Phase I; 8.2 acres are applied to Phase III. | | | |

Source: Panama City- Bay County International Airport (2006b)

## Initial Airfield Development

The proposed initial airfield development is depicted in the rendering below (Please see Figure 2). This project includes the design and construction of a primary runway (Runway 16-34) suitable for commercial operations of narrow-body aircraft with a length of 8,400 feet and a width of 150 feet and a full-length parallel taxiway along the west side of the runway. Taxiways serving Runway 16-34 will be 75 feet wide. Taxiways southwest of Runway 16-34 serving the General Aviation (GA) apron will be 35 feet wide. Taxiways south of Runway 16-34 serving the Runway 3-21 system will be 50 feet wide. Project work also will include construction of a 52,000-square-yard concrete apron surrounding the terminal building.

Throughout the primary runway/taxiway system separation distances, safety areas and object free areas are designed to FAA Airport Reference Code (ARC) D-V standards to maintain the ability to upgrade the taxiways in the future with minimal construction work. The terminal apron and its two connector taxiways are designed to accommodate aircraft in Aircraft Design Group (ADG) III on the south side and ADG IV on the north side. The primary runway system was designed with a 600-foot separation between the runway and the parallel taxiway in order to accommodate standard high-speed exit taxiways in the future.

The crosswind (Runway 3-21) is designed to accommodate operations by a majority of the general aviation fleet and will provide a paved length of 5,000 feet and width of 100 feet. A full-length parallel taxiway will be constructed along the south side of the runway. Taxiways serving Runway 3-21 will be a nominal width of 35 feet; however, select taxiways will be designed with a nominal width of 50 feet to facilitate access of ADG III aircraft to future facilities planned along the southern areas of Runway 3-21.

**Figure 2. Proposed Initial Airfield Development**

The design aircraft for primary runway pavement was identified as the Boeing 767-300ER with a maximum allowable gross takeoff weight of 412,000 pounds and a dual-tandem wheel gear configuration. The pavement design for the crosswind runway will accommodate approximately 6,000 equivalent annual departures by aircraft weighing 70,000 pounds (dual wheel gear). Both runways have been designed with Portland cement concrete (PCC) and asphalt cement concrete (ACC) pavement sections which will be bid as alternates and evaluated on a life-cycle cost basis.

The project also will include the design and construction of a main access road that will connect County Road (CR) 388 to the main terminal area and will initially provide the only access into the airport. This roadway will be a four-lane divided rural controlled-access facility, running north-south for a total length of about 11,600 feet. The access road will have a posted speed limit of 45 miles per hour (mph) outside the airport property, with the speed limit transitioning down to 15 mph as it approaches the terminal.

This roadway will consist of two typical sections. A rural section will run from CR 388 to approximately 800 feet from the terminal area where the typical section will change to an urban section with curb and gutter. The four-lane divided rural section will split into a two-lane, one-way section as the traffic slows so drivers are directed to the parking area or to the loading and unloading areas in front of the terminal.

The GA access road will begin at the wastewater treatment facility, located in the southwest corner of the airport property. It will run along the southerly border of the airport property, and intersect the main access road approximately 1.3 miles north of CR 388. This will be an at-grade, non-signalized intersection. The road will then run about 400 feet to the northeast and then turn northwest to run approximately 3,700 feet. The north end of the GA access road will terminate at the facilities area south of the terminal.

The portion of the roadway west of the main access road is planned to be a four-lane divided road in the future, as is the 400-foot section east of the main access road. The two northern lanes will be constructed in these areas to provide a two-lane roadway (one lane in each direction), with the capability to expand to a divided four-lane road in the future. The 3,700-foot portion of this road will be a two-lane road.

The main access road will transition from a rural section to an urban section at a point approximately 800 feet from the terminal area. At a point 600 feet prior to the transition, the roadway will split to provide one-way passenger traffic to the loading and unloading areas in front of the terminal, or to the parking area. The vehicular traffic nearing the terminal area will be slowed using various traffic calming procedures such as tight curve radii, signing, and landscaping.

Just before the terminal is a taxi staging area adjacent to the terminal loop road. The staging area is composed of two parallel traffic lanes, one for parking and waiting, and the other for circulation. Once a taxi exits the staging area and enters the main access road, it may go back to the terminal area or leave the airport. A taxi waiting facility will be located adjacent to this area for the taxi drivers. A signal will be received at the taxi staging area when a taxi call button is pushed at the terminal curbside.

The airport's main automobile parking area will be located in front (west) of the airport terminal. This area located within the terminal loop road will include public, rental car, and employee parking areas. The public parking area will include approximately 693 parking spaces comprised of a combination of short-term and long-term parking areas. The number of spaces allocated to short-term and long-term parking will be capable of being adjusted by a movable concrete barrier wall. A total of 303 parking spaces will be located within the initial rental car ready lot. An employee parking area with 102 parking spaces will also be located within the terminal loop road.

## Site Grading and Soil Conditions

At the proposed site, high water tables exist in many areas. Dewatering prior to grading and pipe installation will be necessary. In order to minimize earthwork and to maintain the existing drainage pattern from the north to the south, the longitudinal grade of Runway 16-34 site was designed at 0.1 percent, the north end being the high end.

Muck and organic materials were encountered in several borings performed by Environmental & Geotechnical Specialists, Inc. (EGS). The depths of these soils are approximated to range from a minimum of 3 feet to a maximum of 18 feet below grade. Proposed treatment of the muck and organic materials ranged from complete removal in shallow areas and areas of proposed pavement subgrade or structural foundation to surcharge and capping in non-critical structural areas such as stormwater ponds.

The geotechnical report of the existing soil subsurface strata indicated that the existing subgrade California Bearing Ratio (CBR) is low (10-22). Due to the low CBR, the subgrades will be stabilized to depths of 6-12 inches to produce a Load Bearing Range (LBR) of 40 (CBR of 32).

## Drainage Design

The proposed storm water management facilities (see Figure 3 on next page) for the new airport are designed in accordance with drainage requirements of the Federal Aviation Administration (FAA), the Florida Department of Environmental Protection (FDEP) and Bay County. In addition, the storm water treatment systems are consistent with current FAA policy that storm water ponds should not be a wildlife attractant.

Much of the natural drainage within the area has been altered due to silviculture operations. Forest roads and drainage ditches are common and natural drainage features often occur as shallow depression areas with poorly defined drainage ways adjacent to and within the pine plantations. The airport property has several different outfall locations to the south, east, and west. These outfalls ultimately drain to two major systems: Crooked Creek and Burnt Mill Creek, which flow from north to south within well-defined drainage ways that ultimately discharge to the West Bay.

The proposed storm water management system on the airport will consist of storm water collection systems and storm water treatment/attenuation ponds. The collection system will include swales, ditches, inlets, and pipes which collect storm water runoff for conveyance to the storm water ponds. In order to meet FAA wildlife attractant criteria, the storm water treatment facilities are dry ponds with under-drain systems and dry roadside swales. In addition, the airfield conveyance ditches include sub-drainage (under-drain) to aid in the control of the intercepted groundwater table below the ditch bottom.

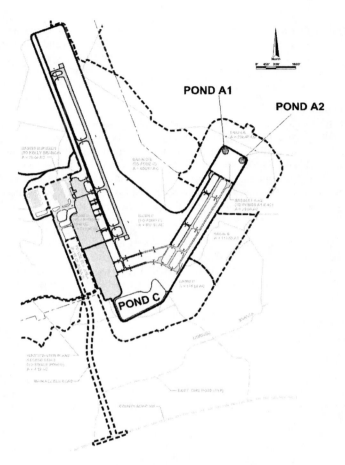

POND A1

POND A2

POND C

**Figure 3. Proposed Stormwater Facilities**

The three major airfield ponds (A1, A2 and C) have been designed to be dry treatment ponds with effluent filtration systems (under-drain). (Please see Figure 4 on following page). The ponds were designed on a voluntary basis to meet the most stringent criteria of the regulatory agencies, plus an additional 50% water quality volume. Recovery of the water quality volume in the treatment ponds will occur within 24 to 48 hours following the storm event.

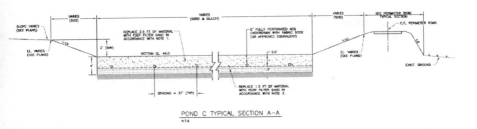

PND C TYPICAL SECTION A-A
N.T.S.

**Figure 4. Proposed Stormwater Ponds**

These storm water facilities meet Bay County Land Development Regulations (LDRs) that limit the post development discharge rates to pre-development discharge rates for the 25-year critical duration storm events at all of the project's outfall locations. The proposed infield ditches and storm water ponds will act as an interconnected system that attenuates storm water before discharging off of airport property.

**Terminal**

The terminal initially will feature three contact gates (with two additional contact gates to be added during the first phase of service) and three non-contact gates (Panama City – Bay County International Airport 2007a, 2007b). It also will allow for a second phase of expansion. Phase I is intended to handle future airport traffic until 2018, at which point the second phase with two additional contact gates and additional apron hard stands will be required. This configuration is anticipated to provide service until 2028. Traffic until the year 2018 is expected at 310,000 annual enplaned passengers. The forecast for 2028 is 398,000 annual enplaned passengers. The common departure lounge will be sized to accommodate six active aircraft. The aircraft stands will accommodate one B767, four B737s and two Regional Jets (RJs).

Facility designs are to be contemporary in character, while still capturing the native Florida Cracker vernacular architecture (a traditional style of building in Florida using wood structures, metal roofs and designed to be passively cooled by the environmental conditions using airy porches and ventilated roofs) allowing the building to contain a historical quality. It will include 101,000 square feet and will have both a single and double height spaces. The first level will accommodate public functions such as check-in, concessions and baggage screening and the second will contain administration offices.

At the outset, it was determined that the terminal was an ideal candidate to incorporate sustainable design principles to meet a Silver Leadership in energy and Environmental Design (LEED) rating, which is a certification of the U.S. Green Building Council under the environmental permitting process.

Under this goal, the building was designed to use natural light and ensure an efficient use of energy. Photo responsive lights also will be utilized to maintain consistent lighting levels along with energy efficient fixtures and lamps. Glazed windows with overhangs will be used to minimize heat gain.

The building also will make use to the maximum extent possible to use recycled and low-emitting materials and renewable resources, including carpeting and seating. Native landscaping will shade and shelter the building and drought tolerant plants will be used. Well water, instead of potable water, will be utilized on site, reducing consumption of fresh water resources.

The building also will be designed to withstand hurricane force winds. The terminal building itself will be able to withstand 130 mph winds, as defined by the National Weather Service.

## Construction Schedule and Cost

Construction of the new airport is slated to begin during summer 2007 with a contract package including site clearing, grading, drainage improvements, paving and airfield lighting. Additional bid packages will follow for master utilities, terminal building, support and general aviation facilities, NAVAIDS, and landscaping. An aggressive 30-month construction schedule is proposed which will result in the opening of the new airport in late 2009.

The total projected cost to relocate the Panama City - Bay County International Airport, as of April 2007, is estimated to be $330 Million. This includes all phases of the project including feasibility and site selection studies, airport master plan, environmental impact statement and mitigation, sector plan, design, permitting and construction, and financing. The projected funding for the project, pending approval of federal grant applications, will be divided between the following sources: the FAA (27%), the State of Florida (36%) and local sources (29%), and a loan from the Florida State Infrastructure bank (8%).

## Conclusion

This project has showcased a rare opportunity to plan and develop a new "green field" airport conforming to current design and environmental standards. Despite a number of geographical, airspace and environmental constraints, the airport development team has successfully navigated the project from concept to groundbreaking in an efficient and responsible manner. Along the way, the airport was able to participate in a partnership with St. Joe to complete the largest Sector Plan ever undertaken in the State of Florida resulting in a 100 year development plan that not only accommodates the new airport, but also provides for the well planned development of over 75,000 acres in north-central Bay County and results in significant amounts of environmental preservation including sensitive watershed and bay shoreline. In order to meet challenges associated with the proposed airport site, such as high groundwater conditions and the fact that nearly half of the 4,000 acre proposed site consists of jurisdictional wetlands, the airport development team has leveraged the opportunity to plan and design mitigation via the restoration of nearly 10,000 acres of industrial pine plantation to the wet pine savanna and flatwoods habitat. Because of the ETP process, an ecological functional assessment of the mitigation plan shows a significant net gain in wetland function relative to wetland impacts. Through the ETP effort, this project sets a precedent for holistic permitting; LEEDS certified buildings, generous and thorough environmental and, storm water facilities, wetland mitigation as well as development of a geographically and socially complementary airport facility from start to finish.

The FAA gave its Record of Decision (ROD) in September 2006, clearing the way for construction to begin in 2007. In December 2006, The Panama City-Bay County Airport (PFN) and Industrial District (Airport Authority) received the final State of Florida permits necessary to move forward with the airport's relocation. Issuance of this state permit has cleared the way for the U.S. Army Corps of Engineers to issue a Section 404 permit, the final permit necessary before construction can begin.

The net result of all these efforts is that the Panama City – Bay County International Airport and Industrial District has successfully met the challenge of addressing the shortcomings of an existing airport which is landlocked, does not meet current FAA standards and has no opportunity for expansion by providing the necessary infrastructure to serve the air service needs of citizens of Bay County and northwest Florida for the next fifty years and beyond.

## Acknowledgements

PBS&J would like to acknowledge the project team that included HTNB, AVCON, IMDC, EGS, and Bechtel which performed the work described in this paper for the Panama City- Bay County International Airport under a Professional Services contract.

## References

Panama City- Bay County International Airport (2007). *Terminal Building Report*, Bechtel Corporation, January 2007.

Panama City- Bay County International Airport /Airport Relocation Project Webpage http://pcairport.bechtel.com/ accessed February 2007.

The St. Joe Company Webpage http://www.joe.com/ accessed May 2007.

Panama City- Bay County International Airport (2006). *Basis of Design Report*, Bechtel Corporation, November 2006.

Panama City- Bay County International Airport (2006). *Final Mitigation Plan*, Bechtel Corporation and PBS&J, October 2006.

# Management Challenges to Redevelopment of a Mature Airport Site: A Case Study of the San Antonio International Airport Expansion Program

Julie E. Kenfield, P.E.,[1] Steven Peters, AIA,[2] and Susan St. Cyr, P.E.[3]

[1] Special Services Manager, PM Team (Carter & Burgess), San Antonio International Airport, 1151 S. Terminal Dr., San Antonio, TX 78216; e-mail: Julie.Kenfield@c-b.com
[2] Program Manager, PM Team (Carter & Burgess), San Antonio International Airport, 1151 S. Terminal Dr., San Antonio, TX 78216; e-mail: Steven.Peters@c-b.com
[3] Sr. Airport Engineer and Expansion Program Development Manager, San Antonio Aviation Department, 9700 Airport Blvd., San Antonio, Texas 78216; e-mail: Susan.StCyr@sanantonio.gov

## Abstract

San Antonio International Airport's $446 million expansion program focused on redevelopment of terminal, roadway, parking, and support facilities to accommodate growth anticipated through 2012, or approximately 5.0 million annual enplanements. Like many mature, urban airports, San Antonio International has grown around its original 1940s era layout to include two terminals, surface and structured parking, and a variety of tenant facilities in the densely developed terminal area. The Airport's master plan addressed options for expansion, recommending replacement of the older 1950s era terminal, extension of the two level terminal roadway, construction of new structured parking, and future expansion of the terminal area to the west. In a schematic design study conducted between 2000 and 2002, the program was validated and defined. In 2003 the City retained two principal A/E teams to design the improvements and a Program Management Team comprised of Carter & Burgess, Inc., Parsons Brinkerhoff, Foster CM Group, and Ricondo & Associates, Inc. The Aviation Department, Program Management Team and the Airport's financial consultants worked to manage the overall budget, contain the effect of rising construction costs, and develop a realistic financing plan. This paper discusses a number of management challenges result from the redevelopment of the "urbanized" airport site. Included in the discussion are highlights such as the extensive contractor and industry outreach for acquiring quality contractor, alternative procurement strategies, and the public outreach program.

## Background

San Antonio International Airport (SAIA) occupies 2,346 acres in north central San Antonio. Developed in a rural setting in the late 1930s, it is now surrounded by commercial, industrial and residential land uses.

The San Antonio Aviation Department manages the aviation operations and facilities of SAIA and Stinson Municipal Airport. The Aviation Director reports to an Assistant City Manager, who oversees aviation and several other city departments. The City Council sets policy and authorizes capital expenditures through approval of ordinances and the annual budget process.

In 2005, SAIA averaged 130 daily commercial flights and handled 7.4 million annual passengers. By mid-2007, 20 airlines will serve 39 domestic and international non-stop destinations with narrow-body and regional aircraft. SAIA is primarily an O&D airport with approximately 11% connecting traffic. It is a designated Port of Entry for U S Customs. From 2003 to 2006, passenger boardings increased by 23%. This growth has been attributed to growth of the San Antonio economy, which substantially increased white collar and manufacturing employment.

### *Facilities*

SAIA has two airline terminals located on the southwest portion of the airfield. Terminal 1, constructed in the early 1980s, and Terminal 2, constructed in the 1950s and expanded in 1967. In all, the terminals have 24 airline gates capable of being served by passenger loading bridges. Through early 2007, three of the gates were not in regular use and were not equipped with bridges. However, by mid-2007, these gates were leased and bridges were installed.

Hourly and daily parking (surface and garage), the central utility plant and the FAA Air Traffic Control Tower and TRACON are located within the terminal roadway loop. Prior to the start of the program, there were approximately 1,323 hourly and 3,537 daily parking spaces. Airport economy parking is located southwest of the terminal area with 1,453 spaces. The Aviation Department operates all of its own parking facilities, providing shuttle service for both employee and economy parking lots. GA facilities surround the terminal area. The terminal area is shown in Figure 1.

### *Program Origin and Development*

To respond to demand and to replace aging facilities, the 1998 Master Plan outlined a program of terminal area development (Ricondo & Associates, Inc. 1998). In 2000, to define terminal and related facility requirements, the Aviation Department retained Gensler with San Antonio A/E firm Marmon Mok to prepare a program document and schematic design. Responsibility for the Expansion Program was assigned to the Planning and Engineering division of the Aviation Department.

The Schematic Design work was delayed by the events of 9/11, but was completed in February 2003. It defined a program, depicted in Figure 2, consisting of two new terminals (B and C), extension of the two-level roadway serving Terminal 1, demolition of Terminal 2, expansion of the central utility plant (CUP), and construction of a new parking garage.

Figure 1:  Airport Terminal Area

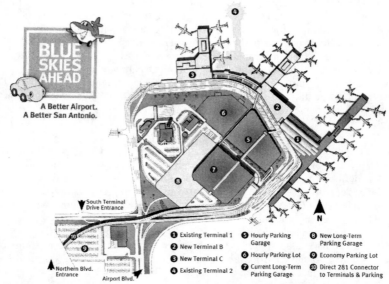

Figure 2 – SAIA Expansion Program and Blue Skies Ahead Logo

To meet forecast demand of 5.98 million annual enplanements (MAE) through 2021, the plan recommended expansion to 34 gates in the three terminals. The Schematic Design Team estimated that 31 gates would satisfy aviation demand through 2015 at the forecast growth. Subsequent environmental analyses allowed for terminal expansion to 28 gates: 16 existing gates in Terminal 1, seven new gates in Terminal B and five new gates in Terminal C. Terminal C could be expanded to 11 gates, but would require additional environmental analysis

The proposed location of Terminal B, between Terminals 1 and 2, required demolition of the east end of Terminal 2 and consolidation of airline activities into about two-thirds of the existing ticketing lobby and bag makeup space. Terminal C could be constructed with only minimal impact to Terminal 2; however, significant apron reconstruction was required for the new terminals due to grading changes. Elevated roadway construction required removal of the curb front canopy and a sliver of the west end of Terminal 2. Existing surface parking would be lost to parking garage construction.

Subsequently, in 2003, the Aviation Department hired the PM Team to implement the program; and two design teams led by 3D/I and HNTB. The program was subdivided into 10 projects, with numerous volumes under each project. The initial projects consisted of utility relocation and construction, and miscellaneous demolition to prepare for construction of the roadway, terminals and parking garage.

## Challenges and Actions

In all projects, lessons are learned as work progresses. In a successful program, such as the SAIA Expansion, processes and methods are altered to improve efficiency, coordination, and final results, to track the program more effectively, and to respond to industry or market changes. The PM Team and Aviation Department have made substantial changes to respond to the market, design challenges, budget, and schedule.

### Project Approach vs. Holistic Approach

The Terminal Programming Study's Schematic Design (Marmon Mok/Gensler 2003) defined development projects within the Expansion Program, including some broad-based utility work to prepare for the terminals and roadway. However, as design and construction has progressed, the PM Team and Aviation Department have adopted a more holistic approach to program implementation. Individual projects in the program have required additional resources and a broader scope than was envisioned. Unanticipated conditions also arose.

One example project is construction of the initial utility relocation and upgrade, addressing replacement of water, sanitary and storm sewer, and electrical and communication duct banks. Construction documents were based on record drawings along with field and subsurface surveys. The documents were technically accurate and included traffic routing around work areas and roadway signage. However, construction revealed that significant pieces were lacking, which created delays, frustrations, and required a rapid response from the PM and Design Teams. Missing pieces included: lack of detailed traffic analysis particular to airport curb front and commercial vehicle operations; inattention to passenger movements throughout the work areas; no consideration for interruption of airline curbfront check-in activity; lack of coordination regarding re-cabling of telephone, communications and FAA duct banks; and inadequate passenger signage inside the terminal and along the curb front to support phasing.  Due to the higher level of customer service expected at the airport, construction conditions generally acceptable on city streets for interim periods are generally unacceptable to the City and Aviation Department.

Through intensive coordination and some redesign during construction, the PM Team addressed these issues alongside the Design Team and Aviation Department.  Although the project was completed with little negative effect on the traveling public, its construction time and contract costs increased, impacting Program and departmental budgets.  This led to a new approach to future projects. The Aviation Department created a weekly coordination meeting that engages the PM Team with the various divisions of the Aviation Department to identify and address construction/operations challenges.  Construction contingencies have been increased to cover customer service initiatives such as additional signage, skycap relocation, and better pedestrian and passenger access.

Design Teams also must take a holistic approach to projects that may be designed separately for bidding, but which are physically adjacent to one another.  At SAIA, the Expansion Program envisioned the completion of Terminal B prior to completion of the adjacent elevated roadway.  This allowed shoring for the terminal foundation to be minimized in favor of laying back the excavation, and provided a location for permanent hydronic (heating and cooling water) lines through crawlspace below the terminal.  However, tenant modifications delayed Terminal B design while the roadway was bid and construction began.  Changes to both projects resulted from revised phasing.  Some of these impacts were not identified prior to bidding, resulting in construction change orders.  The PM Team has the responsibility to support the re-phasing, but cannot be as familiar with the design elements as the members of the Design Team.  So, it is incumbent upon both the designer and PM Team to address the effects of adjacent project changes.

Program implementation affects a wide ranging scope beyond discrete design projects.  At SAIA, additional broad-based work will be required to achieve the upgrade.  These include a landscape plan to unify the look of the plantscape throughout the terminal area and a way finding and signage plan to update and unify terminal, curb front, garage and roadway signage.  Elements of these overall plans have been incorporated into the specific capital projects that they affect (terminals, garage, and roadway) but also must be incorporated into existing facilities such as Terminal 1 and the existing parking garages.  Some of this overarching work was not foreseen by the original program, and was added as work progressed.

### *Engaging Tenants in Design*

A capital program must serve both the airport sponsor, who will continue to own, operate, and maintain the facilities, and the tenants who will conduct business within the facilities.  Therefore, tenants should be engaged in the design of facilities. This is a straightforward – although challenging – process for renovation of existing facilities with known tenants. It is more difficult in new facilities, for which tenants have not been identified.

To understand the needs of the construction project and the operational needs of the tenants, the designers, Owner, PM Team and tenants must communicate. Design teams are selected based on their technical expertise in a specific area. If they focus solely on specific facilities, their design and construction planning will occur in isolation. Contract drawings define the technical aspects of the construction, and often overlook the ongoing operational and customer service needs which complicate design and construction. Only through engagement with airport and tenant staff can these issues successfully illuminate design solutions. Just as designers may be unaware of operational needs, tenants can be equally unaware of complexities of design and impacts of construction. Although it takes time and perseverance for the parties to work through differences, at the end all have a stake in the success of the project.

Tenant involvement may be more difficult for new facilities. San Antonio, like many airports, initiated the Expansion Program based on demand for gates, but without identifying tenants for the new terminals. General discussions of the program with the airlines ignited interest in the new terminals and so the Aviation Department moved ahead. Potential tenants and signatory airlines needed to know the financial impacts of new facilities; for the Aviation Department to calculate those impacts, significant design had to be completed.

One challenge to the Aviation Department was how far to proceed with design before a tenant commitment. In the case of Terminal B, the decision was made to proceed to 90% to confirm costs and design, prior to fully engaging with the airline tenant. This created a need to later modify the design to address needs specific to that tenant. By working through a coordinated floor plan change without engaging in detailed redesign, the PM and Design Teams minimized the financial impacts of redesign. The floor plan change allowed various elements of the terminal to be coordinated with the prospective airline tenant, TSA, and Aviation Department divisions. The revised terminal cost was estimated based upon previously developed estimates of each type of functional space, escalated to the new midpoint of construction. Using this method, redesign costs were minimized and resolution was expedited, while the Aviation Department was provided with cost estimates adequate to initiate lease discussions. Unfortunately, the original tenant decided not to relocate and therefore the process was repeated with new tenants. Terminal C, at approximately 60% design, will undergo a similar process when the need for that facility is confirmed.

### *Addressing Ongoing Change and Maintaining Customer Service*

The one certainty in an expansion program is that change will be continuous for a long time. To maintain efficient operations, the Owner and the PM Team must prepare for those changes. For example, the roadway project at SAIA constructs a new two-level roadway over and adjacent to much of the existing one and two-level roadway. During two years of construction, traffic patterns, passenger flow, and even terminal access will change frequently to permit construction. The "as-designed" phasing required completion of discrete segments of the road working from each end toward the middle of the project. It created an unacceptable reduction in Terminal 2 curb front capacity and pointed out the need to develop centralized commercial ground transportation area outside the roadway footprint but in close proximity to Terminal 2. This was completed prior to award of the roadway contract by another contractor. The PM Team worked with the designers to re-phase the work for bidding, using a similar incremental completion approach, and identifying intermediate milestones at which incentives and liquidated damages would be assessed. This phasing was complicated by the close proximity of a proposed parking structure, which would be under construction concurrently with the roadway. By the time that the contract was awarded, the garage footprint had been reconfigured, and allowed more flexibility for traffic control. The contractor proposed his own version of phasing which would result in a time savings and no additional cost to the owner. However, this phasing was based on construction across the entire roadway footprint for the duration of construction, providing the contractor free access to all areas at the expense of passenger and vehicle access. The PM Team then revised the phasing to reach a compromise that provided better access to the contractor than originally bid, but which also maintained reasonable customer service.

The PM Team must facilitate looking ahead to identify operating constraints; planning with the Owner how to maintain airport functionality. Mitigation opportunities must be identified and either integrated into initial design or added later. At the outset of any program, the costs and implications of change should be recognized and incorporated into the program schedule, work plan and budget. Conditions should be continuously evaluated, and improvements made as warranted by customer service, financial and completion goals. If there is a potential for conflict and delay it is prudent to discuss the possibilities and identify potential actions to resolve budget and schedule challenges that may result.

Continued growth of aviation activity also complicates work. Analyses of passenger and traffic flow and parking demand should be updated to ensure that design and customer service initiatives respond to increased demand, to the extent possible. While the availability of three airline gates at the outset of the San Antonio Program allowed flexibility in moving gate parking and aircraft operations, by mid-2007 these gates were occupied. The ripple effects include changes to apron phasing, aircraft parking, need for temporary facilities, and planning for additional airline relocation. During construction of the partial demolition and tenant relocation to prepare for Terminal B construction, one airline that had been handled by another Terminal 2 airline began an independent operation. This required the PM Team to help the Aviation Department identify space for ticket counters, offices and bag makeup, then coordinate with the airlines and other tenants to work out specifics. The Design Team was given an additional service authorization to quickly prepare the construction documents so that the contractor could prepare the space. The additional design and construction work were funded under contingency.

Tenant development at SAIA has limited the areas available for batch plants, contractor laydown and materials storage. The PM Team's construction trailers will likely be moved by the Aviation Department prior to completion of the program to allow corporate hangar redevelopment. Although tenant revenues generally override non-revenue uses of space, Owners should consider the need for contractor staging and office areas as part of program development. Because business deals change, the PM Team continuously coordinates with Airport Properties to identify and avoid potential conflicts in use.

Active public relations are also vital to communicate change to the general public and airport users. SAIA utilizes its website to inform the public about construction changes, and also has developed direct e-mail that is sent out as needed. To identify upcoming changes and prepare press releases to various groups, the Public Relations division coordinates weekly with the PM Team. A quarterly update video is broadcast regularly on the City's public-access cable channel and is also posted on the Airport's website. Internally, divisions are encouraged to update their staff on the program, because the front line workers field daily questions from travelers. Aviation Department staff and PM Team members also visit civic organizations and other groups to present updates on the program.

Early in the program, the Aviation Department adopted the "Blue Skies Ahead" logo to identify the Expansion Program work. The logo, as shown in Figure 2, has been used on banners and temporary signage throughout the airport to remind travelers that although they may experience delays and frustrations due to construction, they are temporary and will lead to better facilities.

### *Attracting Qualified Contractors in a Strong Construction Market*

When the SAIA Expansion Program was initiated, the economic outlook for the San Antonio region was reasonably good and commercial construction was moderate. Since that time, the economy has been on the upswing with increases in jobs, housing starts, retail and commercial construction and heavy/highway construction. San Antonio is home to a number of federal facilities, which engage ongoing construction programs. Commercial and industrial activity have boomed, due to an influx of jobs and manufacturing. TxDOT has hundreds of miles of new highway construction underway in central Texas. TxDOT's 2006 - 2008 Transportation Improvement Program had over $1.16 billion apportioned and programmed for the San Antonio-Bexar County Metropolitan Area (Texas Department of Transportation 2005). In the wake of Hurricanes Katrina and Rita, and as a result of rebuilding in Afghanistan and Iraq, the San Antonio construction market was facing local, regional, national and international opportunities. In order to attract contractors and subcontractors, the SAIA Expansion Program has had to compete with these other construction demands.

Initial solicitations on the roadway and ramp reconstruction projects produced no bidders. The PM Team met with AGC, ABC and individual contractors to discuss the program and identify the reasons for limited interest. Contractors noted the abundance of other, simpler projects and limited resources with which to pursue work. Compared to other agencies, the City's contracting and payment processes were perceived as bureaucratic. As a result, the PM Team initiated the following:

- Partnering with the Aviation Department and Economic Development Department to hold formal and informal industry and small business outreach sessions. The PM Team briefed interested subcontractors on projects and schedules while the City explained certification and business processes.

- Discussing payment procedures with the Aviation and Finance Departments to improve the review and approval process to expedite the payment process. As a result, contractor payment time has been reduced from 90+ days to 30 days.

- Repackaging of projects to minimize project interdependencies and delays.

- Improving contracting language and terms, including modifying the indemnification clause of construction contracts to terms that were more beneficial to both the City and contractors.

- Incorporating milestone incentives and liquidated damages into the roadway and parking garage projects to help manage schedule and provide financial rewards for early completion.

- Assessing use of alternative project delivery methods, resulting in the use of a Competitive Sealed Proposal process and CM at Risk.

- Restructuring and enhancing the Small Business Economic Development Advocacy (SBEDA) Program to eliminate local requirements on some projects and to emphasize and reward team diversity and small business mentoring.

The resulting renewed contractor interest in the program, along with aggressive solicitation of bidders by the PM Team, yielded multiple bids and proposals on the Roadway, Parking Garage and demolition projects, and six proposals on the CM @ Risk.

### *Alternative Delivery Methods*

To improve contractor response and help ensure quality in the final product, the Aviation Department asked the PM Team to investigate the use of alternative construction delivery methods for vertical construction. A city department, Aviation traditionally used design-bid-build construction delivery. However, the City had used alternative delivery methods on other significant municipal projects, to ensure that bidders were both responsive and qualified. An ordinance enabling selection through Competitive Sealed Proposals (CSP) was in place, and this method was selected for the parking garage (City of San Antonio 2003). The PM Team facilitated review and evaluation of bidders, including interviews. Under CSP, the general contractor (GC) or joint venture submits both qualifications and price, identifying subconsultants. Proposers are evaluated for technical expertise and experience, subconsultant experience, small and minority business participation, and price. Price represents 40% of the evaluation score, with the lowest bidder receiving full points and an algebraic formula used to distribute points to teams with higher bids.

On the SAIA parking garage in June 2006, two bidders submitted proposals that were 33% over the estimated construction cost, but were within less than 1% of each other. The bids were evaluated by the PM Team and presented to the selection panel. The panel interviewed both proposers and chose one as the most responsive and qualified. However, because bids were far over budget, the PM Team performed a technical review and revised the project concept to simplify its design and reduce cost. The Aviation Director and his legal staff used a provision of the CSP process to negotiate the scope and fee of the project with the best qualified proposer (City of San Antonio 2007). The garage design consultant agreed to re-perform the design for reduced compensation and the contractor agreed to provide cost and constructability input for a small fee. The Aviation Director asked the City Council to award the contract at a reduced construction cost, contingent upon redesign to fit the project within that sum.

Through a highly professional, collaborative effort between the designer, the contractor, the Aviation Department and the PM Team, the redesign was completed in mid-December 2006 (City of San Antonio Aviation Department 2006, 2007). To take advantage of an available precasting production slot, the Aviation Department authorized a partial notice to proceed prior to design completion. In mid-January, final pricing was received based on 100% design. The final construction cost, including allowances which may not be expended, was within 2% of the original budget. The project will be completed on the initial schedule, despite time spent on redesign.

Other major construction projects within the Program were to start in 2007. Due to concerns regarding contractor and subcontractor availability in 2007 and 2008, the PM Team helped the Aviation Department strategize on bundling over $108 million of projects under CM at Risk contract, with potential for additional future work of $82 million over a 10-year term. The PM Team and Aviation Department legal staff visited with a number of general contractors (GCs) and with other Texas cities in which this delivery method had been employed. Working with the city's legal staff, the PM Team helped draft an enabling ordinance for City Council consideration, which was approved February 1, 2007. Six qualified firms submitted as GC for this contract, which is scheduled to be in place by July 2007.

Selection of the GC will be qualifications-based, with fee counting for only 5% of the scoring. As in the CSP, the GC with the lowest fee receives all points associated with that evaluation category and the algebraic equation is applied to distribute points to the teams with higher bids. Because the GC will not have subcontractors on board, small/disadvantaged/minority business goal evaluation at selection is based on past performance, anticipated utilization, and the GC's emerging diversity business plan for support and mentoring, including bonding, insurance, and prompt payment programs. For selection, the Aviation Department enlisted a number of non-voting advisory members of the selection panel from airports and airlines who had CM at Risk experience or stakeholder interest. The PM Team facilitated the activities of the selection panel.

Under the CM at Risk contract, overhead and general conditions terms are negotiated after GC selection. These are included in the City Council Action that awards the contract, along with preconstruction services, in this case, on Terminal B and the CUP. As design is completed on individual projects, the GC will bid the subcontracts and buy out the project. Subsequent actions for preconstruction services and construction guaranteed maximum price will include application of the overhead and general conditions equations. Monthly reports will be made to the Aviation Department and City regarding contracted amounts and participation goals.

### Cost, Schedule, and Budget Management

Successful programs need a well defined and thoroughly detailed program budget with adequate contingency to address anticipated definition and unanticipated changes and realism about both hard and soft costs. Since delay almost always translates to increased cost, the Owner should require consultants to check and verify costs throughout programming and design.

Hard costs in the programming stage should include design and construction contingencies appropriate to the stage of design. These may be reduced as design progresses, but construction contingency must remain, especially if the construction contract will provide customer service initiatives, and where work addresses renovation of existing buildings. Inflation and escalation of costs should be anticipated, recognizing that it takes years to move from programming through design and construction.

The PM Team at SAIA developed a detailed Cost Model for the Expansion Program based on its validation of costs in late 2005. Budget items subject to airline negotiation – finish out for voluntary moves, passenger loading bridges, airline club finish out – were shown "below the line". Working in collaboration with the Aviation Department, these project costs were linked to the financing plan so that as projects were designed and implemented, cost changes could be tracked and figures updated. The Cost Model has been modified several times, as significant costs are verified or changed. For instance, bid costs were incorporated with the award of the Roadway, Parking Garage, and demolition projects. In these updates, the costs of other projects were escalated in accordance with the construction schedule. Because updates to the Cost Model are time consuming for both the Aviation Department and the PM Team, formal updates are undertaken only as necessary.

Realistic schedules should be developed early in the program and used to drive deadlines. With the day-to-day pressures of running an airport, the sense of urgency associated with the program schedule tends to wane. The PM Team and Aviation Department must champion attention to schedule by creating and monitoring progress with respect to specific realistic milestones. Unanticipated events which create delays or rephasing of work can be assessed as they arise to determine the impact on budget and schedule, and incorporate required changes. Unplanned setbacks associated with public and private utilities' completion of electric and communications cabling, which were beyond the control of the program, resulted in a delayed start of the Roadway project and delay claims by the contractor.

Time wreaks havoc with decision making. The longer a program is delayed, the greater the probability is that conditions affecting the original decisions will change – resulting in a reevaluation of the original plan. So, when should the Owner step back and reassess the effects of change? In San Antonio this was done in late 2005 and early 2006, approximately 2-1/2 years into the program. Significant passenger growth, changes in anticipated terminal tenants, and dramatic cost escalations suggested that the program phasing should be reevaluated. Scenarios that included rephasing and retention of existing facilities were assessed for capacity enhancement, relative cost, passenger service, construction phasing and operational efficiency. That analysis indicated that the current program should continue. With ongoing growth of passenger traffic and loss of flexibility for airline relocation, alternatives for maintaining aircraft gates in Terminal 2 and increasing efficiency of Terminal 1 may soon be evaluated. In short, the Aviation Department and PM Team should be open to reevaluation – when necessary.

## *Staffing and Management Challenges*

Few owners have the luxury of dedicating all required personnel exclusively to an expansion program. In mid-sized and small airports, staff is expected to continue managing day-to-day responsibilities while overseeing the expansion program. Hiring a program management team to oversee the design work and facilitate implementation can help, provided the reporting relationship allows the team to be effective.

The initial reporting structure at SAIA funneled PM communications and decision-making through the P&E Manager, who was tasked with all communications up the chain of command and with tenants. Separating the PM Team from other Aviation Department divisions and tenants critical the decision-making process degraded communications and slowed progress. Seeking improvement, the Aviation Director changed the structure in late 2005. This empowered the PM Team to effectively manage the program, now reporting daily through a Program Liaison, with regular communications with the Director and senior airport management. To support the program, working alongside the P&E, Legal, and Finance staff and others, the PM Team now assists in preparation of City Council memoranda and financial management. Design decisions are identified by the PM Team and elevated through the Liaison to the Assistant Department Heads and the Director. The Director has clearly delegated responsibility and authority for program decisions to staff members.

## Conclusions

In summary, programs should strive to accurately define budget, scope, and schedule, all of which are dependent variables affecting success. As a program is defined, the airport Owner should engage the various business lines within the organization and tenants in discussion of the program and brainstorming about facilities and operations that will be impacted by implementation. These actions will help ensure that scope is accurately defined initially and user needs are met during program development and implementation. Development and maintenance of an accurate and detailed schedule will support decision-making by reflecting the time required for design, review and approval, procurement, and construction.

The PM Team and Owner must be realistic and comprehensive when defining program costs and impacts to annual operating budgets. The program budget should address costs of specific projects; those associated with overarching improvements needed to support the program; customer service initiatives; contingencies to address unanticipated changes during design and construction; and realistic cost escalation. The Owner should account for department staffing and budget impacts that may be affected by the program.

As SAIA has shown, development of a mature site is not easy, but it is possible through good communications and realistic assessment of budgets, schedule, and scope.

## References

City of San Antonio Aviation Department (2007). *Airport Status Report.* Report to the San Antonio City Council, March 7, 2007.

City of San Antonio Aviation Department (2006). *Excellence in Customer Service Above and Below the Wing, San Antonio Airport System, 2005 Year in Review.* 2006.

Marmon Mok/Gensler (2003). *Terminal Programming Study, San Antonio International Airport, Phase Two – Schematic Design.* February 2003.

Ricondo & Associates, Inc. (1998). *San Antonio International Airport Master Plan Study.* January 1998.

Texas Department of Transportation (2005). *FY 2006 – 2008 Transportation Improvement Program for the San Antonio – Bexar County Metropolitan Area.* (Draft) http://www.dot.state.tx.us/publications/transportation_planning/san_antonio.pdf. April 25, 2005.

City of San Antonio (2003). *Ordinance 98358 "Authorizing the Director of Public Works to make decisions regarding acceptable project types and the appropriate alternative project delivery methods . . ."* October 23, 2003.

City of San Antonio (2007). *Ordinance 2007-02-01-0127 "Expanding the authorization of the Director of Public Works to make decisions regarding acceptable project types and the appropriate alternative delivery method for city projects to include the construction manager at risk alternative delivery method . . ."* February 1, 2007.

# Geotechnical Aspects of the St. Louis-Lambert International Airport Expansion

Kenneth M. Berry, P.E.,[1] Ahmad Hasan, P.E.,[2] and Thomas L. Cooling, P.E.[3]

[1]Senior Project Manager, URS Corporation, 1001 Highlands Plaza Dr, Suite 300, St. Louis, MO 63110; email: kenneth_berry@urscorp.com
[2]Project Manager, Jacobs Engineering, 501 North Broadway, St. Louis, MO 63102; email: Ahmad.Hasan@jacobs.com
[3]Vice President, URS Corporation, 1001 Highlands Plaza Dr, Suite 300, St. Louis, MO 63110; email: tom_cooling@urscorp.com

## Abstract

This paper discusses a major construction project at St. Louis-Lambert International Airport where a new parallel runway with two taxiways was constructed as part of the airport expansion. The paper focuses on the geotechnical aspects of the project, which included time issues (construction expedients), use of chemical additives, slope stability, settlement, and subgrade stabilization. One of the unique challenges of the project was that cuts and fills of up to 30 meters were made as part of the grading for the runway platform and involved moving approximately 12 million cubic meters of soil. In addition, the cut was almost entirely below the groundwater table. The project also involved construction of the first highway tunnel in the state of Missouri.

## Introduction

Design of a new runway for the St. Louis-Lambert International Airport began in 1999. Construction was essentially completed in 2005, and the runway was open to traffic in 2006. The construction included a new runway and two parallel taxiways, the re-routing of several major roads, two large storm water tunnels, and several ancillary facilities. The realignment of Lindbergh Boulevard, a major north-south highway, required a tunnel. The new Lindbergh Tunnel was constructed using cut and cover construction with the new runway and taxiways being placed directly over the tunnel. The overall project involved numerous geotechnical issues including use of chemical additives to allow year round embankment construction, settlement and slope stability of large cuts and fills, and subgrade stabilization. This paper discusses these issues in general. A plan view of the project site is included as Figure 1.

197

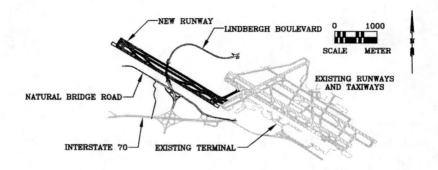

**Figure 1 – Plan view of new runway**

## Construction Expedient

Due to a tight construction schedule, the use of a construction expedient was needed to allow fill placement and compaction year round. Typically, earthwork is not performed in the wintertime in the St. Louis area due to freezing and the difficulty of drying soils to achieve compaction. The project soils contained higher natural moisture contents (5 to 10 percent above the optimum), which required drying to meet the FAA compaction criteria. In the summertime, soils were laid out to dry to reduce the water content prior to compaction. During the wintertime, soils typically do not lose moisture. Moisture from precipitation during the winter aggravated the situation.

In order to facilitate the use of excavated soil as fill, the use of a construction expedient was required. Chemical additives such as Code-L (kiln dust), Class C Flyash, or lime (quicklime) can be used as a construction expedient. A laboratory testing program using the three additives was performed to evaluate the effectiveness of each.

The test program showed that about 2 to 6 percent of Code-L would be needed to improve the workability of the clays in wet weather. The flyash results were highly variable and showed that 5 to 15 percent by weight was needed. Only 3 percent of quicklime was needed. The values stated above include an allowance of 1 percent for loss and uneven mixing during construction. Quicklime provided several benefits. First, it effectively dried the clay to allow it to be compacted immediately. Second, it strengthened the soil mass considerably after curing and reduced its compressibility. This reduced settlement within the fill itself by about 3 times over non-treated clay. The heat generated by curing of the lime also helped reduce the risk of freezing in cold weather.

The site contained primarily modified loess (low plastic clayey silt/silty clay) and residual clays (medium to high plasticity clays). In addition, there were some soft lacustrine soils which consisted mainly of silt. Portions of the shale bedrock were also excavated and then placed in the fill. The shale unit was typical Pennsylvanian- aged shale with interbedded limestone units.

After the study of the quantity of chemical additives was completed, the project team consulted with suppliers of each of these materials to provide the most economical system. The use of quicklime was selected and was added to the soil on an as needed basis. It should be noted that the authors have successfully used all three products on other projects. Site geology, soil chemistry, ambient temperatures during winter months, and economics influenced the selection of quicklime for this project. Strength results of the laboratory tests using quicklime, the selected additive, are summarized in Figure 2.

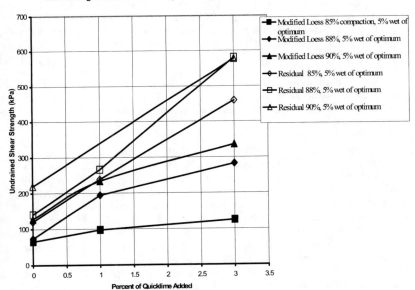

**Figure 2–Results of laboratory tests on compacted samples containing quicklime**

Notes:
1. Lime treated samples were mixed with lime and compacted within 24 hours of mixing. The water content of the samples before adding lime was about 5% wet of optimum. The percent compaction shown in the chart is based on ASTM D1557B (Moist method). After compaction, samples were cured in sealed containers for 7 days. They were then backpressure saturated, consolidated at a confining pressure of 69 kPa, then sheared in undrained triaxial compression at a confining pressure of 69 kPa.
2. Samples with no lime were prepared and tested as noted above, however, they were not cured 7 days before testing.
3. Typical Properties of the untreated modified loess at field moisture were:

      Water content - 24 %          Liquid limit - 34 %
      Plasticity Index - 14 %       Maximum dry density -  18.1 kN/m3
      Optimum Water content -13.5 %

4. Typical Properties of the untreated residual clay at field moisture were:

      Water content - 21 %          Liquid limit - 47 %
      Plasticity Index - 32 %       Maximum dry density -  18.2 kN/m3
      Optimum Water content -14.4 %

Higher percentage of chemical additives was needed to reduce the swell potential of in-situ high plastic clays. This is discussed further in the subgrade stability section of this paper.

During construction, the contractor was required to maintain a supply of quicklime on site so that it could be applied as needed. The contractor used Gators to mix the quicklime into the soil providing a uniform mixture of soil and quicklime. The soil was then compacted. This method worked as anticipated, and allowed the runway embankment fill to be placed and compacted year round.

## Slope Stability

Slope stability was analyzed for both during construction conditions, as well as final design. Large cut slopes up to 25 meters high were made. The cuts encountered a shale unit, which is known locally to have stability issues. The largest cuts were for the installation of the Lindbergh Tunnel. In addition to large cuts on the project, large fills were required for the west end of the runway embankment. There was some existing, undocumented fill in this western area. During the geotechnical exploration phase, some borings and CPT soundings encountered soft creek bed sediments below the undocumented fill. There was a concern about the impact of these materials on embankment settlement and slope stability.

### *Cut Slopes*

There is a history of failure of cut slopes in St. Louis County in areas underlain by shale and its residual clay. Slides commonly occur at the contact of the shale and overlying residual soil, especially when the shale is exposed in the cut. Additionally, these slides are usually triggered by water pressure during periods of wet weather. The key to controlling such slides is adequate drainage, both internally and externally. The tallest cuts on the project, and the ones of most concern, were 20 meter high cuts along the runway, and 25 meter high cuts for the portals for Lindbergh Boulevard. Lindbergh Boulevard was placed in a cut and cover tunnel, which underlies the new airport runway and taxiways.

Temporary cut slopes were recommended to be 1.5H:1V. This included cuts in shale for the construction of the Lindbergh Tunnel. Short term cut slopes in the lacustrine soils were 2H:1V. The temporary cuts in the shale were a concern because of the known history of these deposits. The slope recommendations for permanent cut slopes were made using residual strengths of the shale (friction angle of 16 degrees) instead of peak strengths. This was due to the presence of slickensides encountered in shale samples collected during the drilling of borings for the project and the known problems with this geologic unit.

Early during the construction of the Lindbergh Tunnel, portions of the cut slopes began to ravel and develop shallow slides. As part of the process to investigate the cause of the instability, surveyors were brought to the area to survey the slopes. The survey data indicated that the contractor had graded the slopes to a value closer to 1.2H:1V instead of the desired 1.5H:1V. The contractor then regraded the slopes to the desired angle. No further instabilities were observed.

Permanent cut slopes were recommended to be 3H:1V to facilitate mowing. The target slope stability factors of safety were 1.5 for the static case and 1.1 for the seismic case (10 percent g acceleration). Benches were added as necessary to maintain these targeted factors of safety. In general, cut slopes taller than 21 meters feet in height contained a 9 meter wide bench.

Some minor slope movements have occurred in one stormwater detention basin, which was made by cutting into the shale. The movement occurred as would be expected with the bottom of the failure plane being on top of the shale. Seepage was observed at two locations, and therefore in-situ pore pressures were expected to be higher at these locations. Remediation was made by cutting out the slide, and replacing the soil with a geofabric and gravel. The remedial measures have been in place for over a year at the writing of this paper, and no further signs of distress have been observed.

## Fill Slopes

Target factors of safety for the fill slopes were the same as those used for the cut slopes. Permanent fill slopes were constructed at 3H:1V for all areas where the depth of fill was 15 meters or less without regard to rate of fill placement. For fills greater than 15 meters in height, the calculated factor of safety for short-term conditions was less than desired (~1.3) if the fill was placed in a time period of 12 months or less. The total fill placement occurred over a period of 27 months. Fill was placed in these critical areas over as long a period of time as the project could allow. Issues such as property acquisition restricted the scheduling options available.

In the area with the greatest amount of fill to be placed, there happened to be approximately 6 meters of existing fill placed for a housing subdivision. The geotechnical investigation concluded that the fill was well compacted; however, soft material (slope wash) was left in place prior to that fill placement. The slope wash was left in place as settlement was predicted to be complete prior to paving of the runways. Fill settlement was monitored to confirm this.

Prior to and during fill placement, instrumentation was being monitored to confirm that the slopes were stable. Instrumentation was read on a routine schedule so that increases in pore pressures or deformations could be monitored.

## Instrumentation

In order to confirm adequate stability during construction, instrumentation (e.g. pore pressure sensors, settlement points, inclinometers, toe stakes, and heave points) were installed.

There were two inclinometers which indicated some lateral movement was occurring below grade. The largest movement observed was at an inclinometer installed near the deepest fill. The inclinometer was installed through the soft slope wash that existed below the pre-existing fill. A total of 70 mm of cumulative lateral movement was observed. All but about 5 mm occurred in the first year of embankment placement. The rate of movement decreased and essentially stopped within the next six months. Visual surveys of the fill slope noted no observable slope movement features even a couple years after construction.

The results of the instrumentation program were mixed. Deformation was observed in two of the six inclinometers installed for the large runway embankment fill. The locations were selected to monitor what was thought to be the most critical areas.

Pore pressure readings were made by installing piezometers in the foundation soils below the embankment and wiring to data collection stations outside of the fill placement zone. The main concern with this type of installation is future trenching. Some wires were cut by trenching after installation.

Settlement gages and heave points rose through the fills and cuts, and were subjected to damage by the contractor. Some redundancy was built into the system so that the overall monitoring confirmed the design assumptions, and satisfied the designers that the construction was proceeding satisfactory.

## Settlement

Settlement was another aspect of the project monitored by the geotechnical designers. Part of the construction sequencing was based upon settlement issues. Differential settlement was a concern for the runway pavement designer. Therefore a small surcharge was placed to limit post construction settlement.

Analyses indicated that primary consolidation was to be completed in about 6 months after placement of the fill. After that time, secondary compression was expected. In 50 years, secondary compression was expected to cause another 50 mm of total settlement after the runway was complete in the area of thickest fill. Differential settlement will be much less.

Settlement of fills in other portions of the runway and taxiway embankment were anticipated to be less due to less fill placed and differences in existing subsurface conditions (eg. lesser thickness of compressible soils). Since settlement was expected to occur mainly in the first 6 months after completion of fill placement, paving operations (placement of the concrete) did not begin until after this time had passed. Settlement gages and surveying within the area confirmed these estimates.

It was anticipated that the maximum settlement of the runway embankment would be in the range of 600 to 900 mm for non-ravine and ravine areas, respectively. Ravine areas were the areas were slopewash was below the pre-existing fills. The total settlement included primary, secondary, and settlement due to self-weight of the new fill. The settlement anticipated 6 months after fill placement ranged from 550 to 840 mm.

Settlement gages indicated maximum settlements during construction in the range of 150 to 300 mm. This was considerably less than anticipated. The reasons for this are twofold. Some of the instrumentation was damaged by the contractor. The amount of settlement that occurred in the time while the instrument was awaiting repairs is unknown. It sometimes took months before an instrument was repaired.

Also, the quantity of settlement that occurred within the embankment fill itself was drastically less than estimated. Settlement within the fill was estimated based on consolidation tests for non-lime treated soil. As it turned out, much of the embankment fill was lime treated to facilitate fill placement. As a result, the settlement within the fill itself was about 1/3 of that anticipated.

## Subgrade Stabilization

The condition of the subgrade was variable across the site. A portion of the site contained soft lakebed deposits. This material was removed and replaced. Geotextile fabric was used to stabilize the subgrade and for use as a separator in areas where the remove and replace technique was employed.

Geofabric was used in selected weaker subgrade areas of the runway/taxiways. In general, approximately 600 to 1200 mm of subgrade was removed. A heavy woven geofabric was placed on reasonably smooth area free of mounds, debris or projections. The geofabric was overlapped such that 100 percent overlap was achieved. This resulted in two layers of geofabric being placed. A material consisting of a gradation of approximately 100 mm minus material was then placed in 300 mm lifts. This process was performed in localized areas where soft subgrade was encountered.

Shale exposed at the subgrade level was over excavated and replaced. The shale was undercut due to its swell potential as well as an attempt to have a more uniform stiffness for the overall subgrade. The shale was over excavated by 1 meter, and replaced with compacted low plastic clay.

It was recommended that high plasticity clays exposed in cuts be chemically treated in place due to the swell potential of these materials. The percentage of chemicals for additives was evaluated in the same study as that for construction expediency. It was determined that 11 percent Code-L or 6 percent quicklime was needed. Up to 15 percent flyash was used in the preproduction testing, but the samples did not achieve the desired swell reduction. Therefore, the use of flyash was not recommended for addressing swell potential at this site. As an alternate to chemical treatment, removal and replacement of high plastic clay similar to the shale was allowed. In general, the high plastic clay was overexcavated similar to the shale. During construction, the top 300 mm of the subgrade was treated using quicklime to form a consistent stable platform for construction of the runway pavement.

Plate load tests were performed to verify the subgrade modulus. Three locations were selected. One test was performed in the large fill area at the western part of the site. One test was performed in the shale cut area in the central part of the site. One test was performed at the eastern part of the site where lacustrine soils were encountered. The pavements for the project were designed using a modulus of subgrade reaction of 100 pci. The plate load tests confirmed the use of this value.

## Closure

The majority of fill placement occurred during 2004 and 2005. The runway pavement was placed in 2005 and the runway was opened for commercial traffic in 2006. The earthwork contractor for the project was Dave Kolb Grading of St. Louis.

The overall project was completed on time and under budget. Given the complexities of the project, the project team is proud of this accomplishment. To our knowledge, the issues that arose during construction are the ones stated herein. There are no open issues, and all geotechnical issues have successfully been addressed. The design and behavior of the various geotechnical items were as anticipated.

Two of the authors worked on the geotechnical engineering design. The other author worked for the program manager. It was extremely beneficial for the design engineers to be able to work with someone on the program management team who was familiar with the design issues involved.

# Reconstruction of Runway 10-28 and Taxiway "H" at Luis Muñoz Marin International Airport, Carolina, Puerto Rico

## Carlos J. Arboleda-Osorio, P.E., M.ASCE[1]

[1]Program Manager, PBS&J Caribe Engineering, CSP, 268 Muñoz Rivera Ave., Suite 1602, Western Bank World Plaza, San Juan, Puerto Rico 00918; PH (787)294-2010; FAX (787) 294-2002; email: cjarboleda@pbsj.com

**Abstract**

Runway 10-28, which runs parallel to Taxiway H, and connector taxiways at Luis Muñoz Marín International Airport (LMMIA) were originally constructed in 1971. The rehabilitation of Runway 10-28 and associated taxiway system involved upgrading the capacity of the pavement system to handle the growing traffic volume at LMMIA. The existing runway and taxiway pavements were significantly past the original 20-year design life and were showing extensive signs of overstress due to growing traffic volume and to an increasing number of wide-body aircrafts operating at the airport. Following the pavement design analysis, the Owner accepted the total reconstruction alternative instead of partial rehabilitation (slab by slab) of the pavement. The total reconstruction of the pavement system included the rubblization of Runway 10-28 and parts of Taxiway H, a leveling course and a new 16-inch (406 mm) Portland Cement Concrete (PCC) overlay. All the connector taxiways and the middle portion of Taxiway H included full depth repair with a combination of permeable drainage layer placed over a 4-inch (101 mm) P-401 asphalt base course and an underdrain system installed on the outer edges. Transition from the new taxiway elevations to the surrounding apron areas took place through full-depth reconstruction in those segments. This alternative removed the causes of the distresses, provided a stable construction platform (rubblized PCC), and provided a full 20-year life to the project pavements.

## Introduction

Luis Muñoz Marin International Airport (LMMIA) in Carolina, Puerto Rico, is the gateway to the Caribbean. Runway 10-28, Taxiway H, and connector taxiways, were originally built in 1971. The reconstruction of the 8,016 ft (2,438 m) long Runway 10-28 and associated taxiway system involved upgrading the capacity of the pavement system to handle the growing traffic volume. The existing runway and taxiway pavements were significantly past the original 20-year design life and were showing extensive signs of overstress due to growing traffic volume and to an increasing number of wide-body aircrafts operating at the airport. Figure 1 shows an aerial picture of LMMIA.

The original runway and taxiway pavement section consisted of 14-inch (355 mm) PCC pavement with a 6-inch (152 mm) soil-cement subbase. The 1,000 ft (305 m) western portion of the runway end had 15-inch (381 mm) PCC for the departures. As part of the investigative phase, a nondestructive testing program of 504 test points was performed by Applied Research Associates, Inc. (formerly ERES Consultants) utilizing a Dynatest Model 8081 Heavy Weight Deflectometer (HWD). Several rehabilitation methods were considered for the runway and taxiway pavements. The following alternatives were included in the pavement design analysis:

1. Slab replacement, spall repair, crack sealing, and joint resealing
2. Runway 10-28 keel replacement, partial reconstruction of Taxiway H, and slab replacement for the remaining taxiway pavements
3. Reconstruction of Runway 10-28, partial reconstruction of Taxiway H, and slab replacements for the remaining taxiway pavements
4. Rubblize Runway 10-28 and Taxiway H and overlay with PCC and full depth repair for connector taxiways
5. Unbonded PCC overlay on Runway 10-28 and Taxiway H and full depth repair for connector taxiways

Of the various alternatives considered for Runway 10-28 and Taxiway H, rubblization of the existing PCC pavement and placement of a PCC overlay was recommended. This alternative had the best combination of initial cost, acceptable life cycle costs, and a shorter construction duration using the rubblized layer as a stable construction platform. Likewise, this same alternative was recommended for the connector taxiways. However, some portions of Taxiway H and the connector taxiways were reconstructed with a new PCC pavement to allow transition of the overlay grades to existing taxiways and apron areas.

**Data Collection**

PBS&J conducted an extensive pavement and geotechnical investigation in November 2002. Data collected included a Federal Aviation Administration (FAA) Pavement Condition Index (PCI) distress survey, a detailed slab-by-slab distress survey, Heavy Weight Deflectometer (HWD) nondestructive testing, geotechnical cores and borings, and a topographic survey. The intent of the data collection was to characterize the existing pavement properties, to develop a pavement management system consistent with FAA's Advisory Circular 150/5380-6, and to develop construction documents for pavement rehabilitation on a slab-by-slab basis.

The pavement surfaces at LMMIA were visually inspected using the Pavement Condition Index (PCI) procedure as outlined in ASTM (American Society of Testing Materials) D5340. The PCI procedure has become the universally accepted means of rating airfield pavements for operational condition and structural integrity. Prior to conducting a PCI inspection, the LMMIA pavement network was divided into the pavement hierarchy of sites, facilities, sections, and sample units. Sample units represent the level at which the PCI inspection was conducted and were typically $20 \pm 8$ slabs for PCC pavements. In addition to the PCI distress survey, a detailed slab-by-slab distress survey was conducted of 100 percent of the pavement. That distress survey identified which repairs were needed at every slab. The major distresses were joint spalls, corner spalls, cracked slabs, shattered slabs, faulting, and surface erosion of the concrete pavement.

The Dynatest Model 8081 HWD was used by Applied Research Associates, Inc. (ARA) to conduct non-destructive testing. The HWD is an impact device that simulates the effects of a heavy, moving aircraft load by dropping a weight over a circular loading plate placed on the pavement surface. The resultant surface deflections are measured by velocity transducers spaced at given intervals in a line extending from the center of the loading plate forward. The HWD is trailer-mounted and operated from a computer mounted in the tow vehicle. Equipment used by ARA is shown in Figure 2.

ARA analyzed the HWD load and deflection data in conjunction with pavement layer thicknesses to determine its structural characteristics, including pavement layer and subgrade moduli. ARA used its in-house deflection analysis software, *Deflexus*. In addition to layer moduli backcalculation, the transverse and longitudinal joint load transfer efficiency as well as a loss of support analysis for the concrete slabs were calculated. Based on the results of load transfer between slabs it was determined that key way failures were present on the longitudinal construction joints in the keel.

The geotechnical investigation was conducted by Geo-Engineering, Inc., which included cores and borings. A total of 45 cores and 20 borings, to a depth of 6 feet (1.8 m), were conducted. Major tests conducted included layer thickness, Standard Penetration Blow Counts (N-value), In-place Moisture-Density at various depths and Atterberg Limits. Generally, the top 2 feet of the runway subgrade is classified as a fill material, which is a silty, sandy mixture. It has a natural water content of 11 to 28 percent and standard penetration blow counts of 58 to 98, averaging 68. The unconfined compressive strength of the fill layer averages 49 psi (3.44 kg/cm2).

**Final Design**

What began as a typical slab replacement project ended in a full reconstruction due to the severity of the distresses and the lack of load-transfer capabilities of the pavements. The runway distresses encountered were a result of the combination of slab geometry, unstable subgrade soils, and the keyways. Current FAA standards require a stabilized base be placed beneath new concrete pavements. Runway 10-28 did not have a stabilized layer, and the subgrade soil provided non-uniform support. Slab sizes of 20 by 25 feet (6.1 by 7.6 m) had a slab aspect ratio of 1.25 to 1. This ratio is considered critical thus generating greater curl stresses in the long direction and inducing cracks at midpoints. The keyways failed in many areas and load transfer was less than 50%, especially on the keel section of the runway. These factors caused the cracking and spalling, which were generating the foreign object damage (FOD). The combination of loss of load transfer and large slab size and aspect ratios contributed to the majority of slab failures on the runway.

Based on the findings the Owner, Puerto Rico Port Authority (PRPA), instructed Post, Buckley, Schuh, & Jernigan (PBS&J) to proceed with the design alternative that would yield the best long-term solution and would guarantee that the new pavement would last at least 20 to 25 years without incurring major maintenance costs. Among the alternatives presented were: slab by slab repairs; keel reconstruction and slab repairs; runway and taxiway reconstruction using full depth reconstruction; runway and taxiway reconstruction using rubblization with PCC surface; and runway and taxiway reconstruction using unbonded PCC overlay. The alternative selected was the rubblization of the existing runway and the construction of the new PCC surface.

This option utilized the existing pavement by rubblizing it into a high quality base, as shown in Figure 3. The Rubblizing technique is new and unique to Puerto Rico, which transformed the concrete into a 3-inch (76 mm) minus fractured concrete particles at the surface and 9-inch (228 mm) concrete particles in the bottom half. For design purposes, the 14-inch (355 mm) rubblized PCC section is better than a high-quality crushed aggregate P-209 base having elastic properties (modulus) in the order of 300,000 psi versus 40,000 psi of the crushed aggregate base. Due to the rubblization process being in place, the layer meets the FAA guidelines for a stabilized base, because the layer has very few fines and is extremely well interlocked. A new standard specification for rubblization was developed in accordance to FAA guidelines, P-215. The FAA reviewed this specification and approved it for this project. The surface of the rubblized layer received a nominal 4 to 6 inch (101 to 152 mm) leveling course consisting of crushed aggregate. The new pavement section for both, runway and taxiways is a 16-inch (406 mm) PCC section, using the 18.75 by 20 ft (5.7 by 6.1 m) slab dimensions. The drainage layer was placed on a 4-inch (101 mm) P-401 asphalt base course. Although a stabilized drainage layer (Specification P-305) was considered as a stabilized base, it was recommended that the P-401 base be utilized to provide a solid support for the drainage layer and concrete. Figure 4 shows the pavement sections for Runway 10-28 and Taxiway H rubblization and for Taxiway H reconstruction.

This option was expected to perform better than the full depth reconstruction option because the existing subgrade would not be exposed to the rain and the high water table, and the rubblized section was expected to provide better long-term uniform support than the gravel and Hot Mix Asphalt (HMA) base combination of the full depth sections due to better aggregate interlock.

## Phasing and Coordination

A particular and key element to this pavement reconstruction was the construction phasing plan and proper coordination with all the key players at LMMIA (FAA Air Traffic Control Tower, PRPA Operations and Management staff, users, pilots, and cargo tenants). Several meetings with the Traffic Control manager helped PBS&J to understand the logistics of the ground movement activities at LMMIA as well as quantify the operational impacts to the incoming traffic at final approach. Allowing all construction work to have been accomplished simultaneously would have had a major impact on LMMIA operations, which would have been unacceptable. Therefore, a detailed construction phasing was developed to lessen the operational impacts. Under the phasing plan Runway 10-28 Reconstruction and Taxiway H Reconstruction were accomplished in several phases to limit the time of runway closure and to provide taxiway access to facilities during the project. Figure 5 shows the latest phasing plan currently been used by the contractor.

## Bidding Phase

Due to the complexity of the work involved and the extensive coordination and phasing activities, the bidding process of this project was extended beyond the actual dates envisioned. This project went through two bidding processes and in each of the processes only one bidder submitted an offer. Both offers were above the engineer's estimate by at least 42 percent. After opening of the second bid, the FAA granted PRPA the option of going to a sole source option and provided the opportunity to negotiate with qualified contractors whose expertise on the paving industry guaranteed the quality work requested for this project.

The international firm of LAGAN International, based in the UK, in a joint venture with the local firm of RBR Construction was selected to perform the reconstruction of Runway 10-28. A final price was agreed upon between the Owner and contractor in the amount of $61.8M. Notice to proceed was finally issued in April 2005.

## Construction

Overall construction activities began in late April 2005. One major activity that was incidental to this reconstruction was the pavement evaluation and subsequent overlay of the primary runway, 8-26, before the contractor started the closing of the Runway 10-28. During the coordination meeting with FAA and PRPA it was necessary to evaluate the structural condition of the primary runway to ensure that no failures occurred during the closure period. Contractor was authorized to do this work as a change order to the project. With LMMIA having only two runways at the time, losing runway 8-26 would have require diverting traffic to other airports and caused major delays to the airline industry and the public.

Evaluation protocol that included a visual distress survey, Pavement Condition Index (PCI), ground penetrating radar (GPR) and heavy weight deflectometer (HWD) testing was developed to identify the immediate and near-future pavement needs in order to implement a proactive repair plan. Pavement evaluation consisted in fieldwork testing including geotechnical sampling and recommendations. The contractor completed the reconstruction on Runway 10-28 during the summer of 2005.

During the months prior to the closure of the Runway 10-28, the contractor procured and installed a new concrete central batching plant on airport premises. The batch plant erected is presented in Figure 6. Because of the limited availability of real estate at the airport, a small 3-acre (1.2 ha) parcel was used for the installation of the batch plant and for the stockpile of the aggregates. Essential for the quality control of the concrete to be placed at the pavements was the hauling times between the central batch plant and the runway, which could not be more than 30 minutes. The contractor prepared a test section consisting of two paving lanes. In this test section the contractor demonstrated the proposed techniques of mixing, hauling, placing, consolidating, finishing, curing, start-up procedures, testing methods, plant operations, and the preparation of the construction joints. Also, during this period the contractor began the demolition work of part of Taxiway H (Phase I). This initial phase set the "learning curve" for the contractor to position his labor and equipment in an area not as critical as the runway. Initial problems on the correct combination and gradations of aggregates resulted on concrete that was not properly consolidated. Once the contractor's paving crews knew the proper placing and finishing steps and the concrete mix designs were adjusted on the batch plant, the contractor was ready to start Phases II and III at the runway.

Another of the main activities performed was the installation of a new storm sewer drainage system. The existing storm drainage system between Runway 10-28 and Taxiway H had inlet structures located inside of the runway safety area and three outfall pipes that crossed under the runway. Since this drainage system was constructed approximately 30 years ago and had pipes under the runway pavement it was recommended to demolish this old system.

The new replacement drainage system has all inlet structures outside of the runway and taxiway safety areas. The location and elevations of the new inlet were designed to be compatible with the new runway and taxiway grades and the new taxiway geometry. The outfall pipes for this system cross under Taxiway H and H-1 instead of under the runway, which reduces any potential problems to the pavements in the future.

First partial runway closure was performed on November 2005 (Phase II) when the first 3,100 ft (945 m)of Runway 10-28 was closed to begin the rubblization activities. The remainder of the runway, 4,300 ft (1,310 m) was kept open for departures of aircrafts that are categorized as Groups I and II. Two weeks prior to this closure the contractor began the installation of the underdrain system parallel to the runway. This underdrain system helps to dissipate water pore pressure generated by the rubblization equipment and remove all runoff and perched water that may be trapped below the pavement to appropriate storm sewer systems.

After rubblization and paving activities of Phase III, the contractor proceeded with the full closure of the runway 10-28 on February 2006 (Phase IV). During this phase the contractor finished the rubblization of the rest of the runway leaving a 200 ft (61 m) threshold at the end of the runway to allow traffic from one of the cargo tenants to access the operational areas of the airport. Construction phasing plans for phases V through VII were altered to accommodate unforeseen conditions such as the removal and stabilization of unstable subsoil material encountered underneath connecting taxiways H5, H6, and H9 as well as under the new quadruple storm drainage system. In addition, the availability of construction areas was limited by the airport management requiring further subdivision of the phases, hence, extending the construction time of the project.

As of January 2007, the contractor completed 7,800 ft (2,377 m) of the 8,016 ft (2,443 m) in length of Runway 10-28 in addition to completing the reconstruction of connecting Taxiways H1, H2, H3, H5, H6, and H9. The contractor completed the subbase installation of Taxiway H where a transition from rubblized to full reconstruction was needed in order to correct the transverse grade to shift the crownpoint to the centerline and match the longitudinal grade. Figure 7 shows the actual progress as of April 27, 2007.

The remainder of work to be done involves concrete removal and replacement of various connecting taxiways to the cargo apron and to the main terminal ramps. Because of the operational constraints, this work will take additional time with substantial completion of the project estimated to be November 2007.

## Conclusion

After a thorough investigation process, PBS&J was able to prepare a pavement design that gave the best long-term solution to the Owner and guaranteed a life expectancy of 20 to 25 years. The coordination with all end-users proves a very useful tool for the construction phasing of this project. Yet, trying to please everyone is something that cannot be achieved and sometimes becomes detrimental in the proper execution of the construction activities. A full runway closure would have had the effect of reducing at least 3 to 4 months on the overall schedule. More detailed information to the users during the design process will be needed in order to have the assurance that everyone will know the implications of the construction activities before they begin.

## Acknowledgements

The author would like to acknowledge PBS&J engineers William Stamper and Dale Stubbs for their design efforts on the project reported in this paper. The author also would like to acknowledge the work of engineer John R. Anderson, Ph.D., from Tigerbrain Engineering in the coordination, construction phasing, and pavement evaluation of Runway 10-28. Other project personnel and consulting firms that collaborated on the initial studies, design, and construction are as follows: Doug Steele and William Weiss from Applied Research Associates, Inc., who were responsible for the HWD testing and analyses of load and deflection data; Armando Rovira from FAA Orlando Airport District Office; Ivonne Toledo from FAA Air Traffic Control Tower; Fred Sosa and Felix Rivera from PRPA Aviation Division; Ildefonso Salva, construction manager from LMC & Associates; Donald Bloodworth, contractor from LAGAN Puerto Rico Limited and Rafael Betances from RBR Construction. The assistance and collaboration between designers, owner, the contractor, and construction managers were vital to building a successful relationship that ensured a seamless project for the benefit of the end users, the people of Puerto Rico and visitors to the island.

## LIST OF FIGURES

**Figure 1. Aerial photo of Luis Muñoz Marin International Airport –
Carolina,    Puerto Rico**

**Figure 2.  HWD Equipment testing pavement at Taxiway H**

**Figure 3. Antigo Construction rubblization equipment fracturing pavement at Runway 10-28**

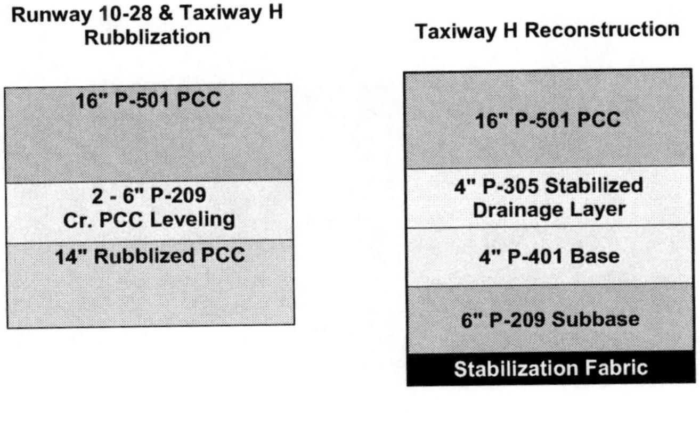

Runway 10-28 & Taxiway H
Rubblization

| 16" P-501 PCC |
| 2 - 6" P-209 Cr. PCC Leveling |
| 14" Rubblized PCC |

Taxiway H Reconstruction

| 16" P-501 PCC |
| 4" P-305 Stabilized Drainage Layer |
| 4" P-401 Base |
| 6" P-209 Subbase |
| Stabilization Fabric |

**Figure 4. Pavement sections for Runway 10-28 and Taxiway H rubblization and for Taxiway H reconstruction**

| RUNWAY 10-28 RECONSTRUCTION SCHEDULE | | | | |
|---|---|---|---|---|
| Project Description | Estimated Start Date | Actual Start Date | Estimated Completion Date | Actual Completion Date |
| Reconstruction Runway 10/28 - Includes Taxiway "H" and Connector Taxiways | | | | |
| > Phase 1 | | 05/23/05 | 12/01/05 | |
| > Phase 2 | | 10/31/05 | | 11/04/05 |
| > Phase 3 | | 10/31/05 | | 04/20/06 |
| > Phases 4 & 5 | 02/06/06 | 04/07/06 | 04/30/07 | |
| > Phase 6a | | 01/15/07 | 05/31/07 | |
| > Phase 6b | 07/15/07 | | 10/31/07 | |
| > Phase 7a | | 04/21/06 | 04/30/07 | |
| > Phase 7b | 02/01/07 | | 07/31/07 | |
| > Phase 7c | | 01/15/07 | 05/31/07 | |
| > Phase 7d | | 08/19/06 | 07/31/07 | |
| | | | | |
| All Phases (Construction) Completed | | | 11/30/07 | |
| | | | | |

**Figure 5.  Construction Phasing**

**Figure 6.  Central Batch Plant installed in project premises**

**Figure 7.** Aerial photo of Runway 10-28, April 27, 2007 (photo provided by Lagan Puerto Rico Limited – used by permission)

# Performance Prediction Of Contingency Airfields Subjected To Repetitive C-17 Loadings

E. Heymsfield,[1] J.F. Peters,[2] and R.E. Wahl[3]

[1]University of Arkansas, Department of Civil Engineering, 4160 Bell Engineering Center, Fayetteville, Arkansas 72701; PH (479) 575-7586; FAX (479) 575-7168; email: ernie@uark.edu

[2]U.S. Army Engineer Research and Development Center, Geotechnical and Structures Laboratory, Airfields and Pavements Branch, 3909 Halls Ferry Road, Vicksburg, MS 39180; PH (601) 634-2590; FAX (601) 634-3020; email: petersj@wes.army.mil

[3]U.S. Army Engineer Research and Development Center, Geotechnical and Structures Laboratory, Airfields and Pavements Branch, 3909 Halls Ferry Road, Vicksburg, MS 39180; PH (601) 634-3632; FAX (601) 634-3020; email: wahlr@wes.army.mil

## Abstract

The U.S. Army Engineering Research and Development Center is investigating methods to construct a contingency airfield for C-17 aircraft loadings within a 72-hour construction period. The proposed method to satisfy these stringent design requirements is to use a chemically stabilized in-situ top soil layer as a runway surface. A suite of soil-chemical mixture combinations was examined to optimize the stabilized soil mixture behavior. A numerical approach is developed in this study to supplement limited C-17 full-scale testing data. Three stabilized soil mixture types are investigated using as stabilizing agents: 6% cement, 4% cement, and 4% cement + 1.5-in. (38-mm) long polypropylene fibers. Unconfined compression tests under monotonic loading and unconfined compression tests using repeated load tests were conducted to evaluate the stabilized soil mixture for strength and durability. Results have shown that the stabilized soil mixture exhibits damage and plastic behavior. In this study, the Army Corps' finite element code STUBBS is enhanced by including material damage to predict stabilized soil behavior when subjected to repetitive loadings. Damage behavior in the Valanis numerical model is calibrated using only two material properties and two damage parameters. Preliminary work on incorporating plastic behavior to supplement damage is discussed. In earlier work, forces induced by a C-17 aircraft on a runway were limited to a vertical force. However, C-17 aircraft braking forces induced during landing are substantial. Therefore, in this work, in addition to the vertical pressure exerted by the aircraft wheels, preliminary work is presented which considers an aircraft braking force.

## Introduction

The U.S. Army Corps of Engineers has been tasked to construct in-theatre contingency airfields within limited time constraints to support C-17 aircraft loadings. Construction of the airfields is to be performed using minimum manpower and equipment. This ongoing research work has been conducted under the US Army Corps of Engineers' Joint Rapid Airfield Construction Program (JRAC). In order to satisfy these stringent requirements, the Army Corps of Engineers have concluded to create the runway surface using the site's top soil layer and stabilizing it with a soil-chemical mixture. Because of the extensiveness of the problem, a numerical approach was desired to supplement already available full-scale test results.

Stabilized soil has a semi-brittle behavior. Therefore, to represent this behavior a numerical model based on damage was considered. Specifically, the damage model developed by Valanis (Valanis 1988) was implemented because of its ease to calibrate. The damage model was incorporated into the Army Corps' STUBBS finite element code so as to evaluate the performance of a stabilized soil runway surface. The finite element code STUBBS was developed by J.F. Peters at the U.S. Army Engineering Research and Development Center (ERDC) (Peters et al. 1997). A description of the code is given in the reference by Heymsfield (Heymsfield 2006).

The following sections of this article review the formulation of the Valanis model and its suitability to model stabilized soil behavior. Three stabilized soil types were studied, soil stabilized with 6% cement, 4% cement, and 4% cement with 1.5-in. polypropylene fibers.

## Background

Airfields using either flexible (asphalt cement concrete) or a rigid (Portland cement concrete) pavement are designed by the Army Corps of Engineers using a semi-empirical approach (Huang 2004; UFC 2001). Stresses induced by the aircraft are calculated using Westergaard equations for rigid pavement runways or the layered elastic method for flexible pavements. The calculated stresses are then converted to number of aircraft coverages based on empirical relationships developed during World War II full-scale testing experiments. Although a design approach is available for flexible and rigid pavement runway surfaces, a design approach for contingency runways constructed using stabilized soil is nonexistent.

Previously, non-linear elastic-plastic behavior was incorporated into the STUBBS code through the multi-mechanical model (Peters 1997). However, using a non-linear elastic-plastic behavior model requires substantial computation time. Instead, because of the brittle behavior of stabilized soil, a damage model was instead used. Other finite elements codes exist which consider nonlinear material properties, ILLI-PAVE (Raad and Figueroa 1980) and MICH-PAVE (Harichandran et al. 1989). These codes assume a hardened surface supported by structural layers and determine material degradation through a transfer function. Conversely, a direct relationship exists between the damage model and material degradation through the integrity tensor. The Valanis damage model was incorporated in this study because of its ease to calibrate; however, since it was developed for a brittle or semi-brittle material, plastic behavior was not considered. Although the predominant behavior of soil stabilized with cement is damage, the authors found in this study that plastic behavior does exist. An approximate approach to complement plastic behavior with damage is discussed in this article.

**Stabilized Soil Material**

Three stabilized soil mixtures were evaluated in this study to establish an optimal stabilizing agent. The non-plastic silty sand soil material consists of 84% river sand and 16% Vicksburg silt. The soil material was then stabilized using 4% and 6% Type III Portland Cement, a high early strength cement, for two of the samples, and 4% Type III cement and 1.5-in polypropylene fibers in a third sample. Water was added to the mixtures to attain a 7.7% moisture content.

A total of eighty-one unconfined compression tests were performed on these samples to examine material behavior. The unconfined compression test is easy to perform and provides a means damage model calibration since a damage model analytic solution exists for simple axial compression. During the tests, radial deformeters measured radial displacement and two LVDT's measured vertical displacement. A load cell concurrently measured applied loads, Figure 1. These measurements were then converted to strain and stress values to examine material behavior.

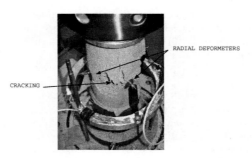

**Figure 1. Unconfined Compression Testing**

**Valanis Model Formulation**

The damage model developed by Valanis (Valanis, 1988) was selected in this study because of its intrinsic characteristics to:

- predict crack development due to incremental tensile strains in tensile strain regions, and
- be calibrated using two elastic material properties and two calibration parameters.

For an elastic material, the constitutive relationship between stress and strain is:

$$s_{mn} = d_{mn} 1 \epsilon_{kk} + 2m \epsilon_{mn} \tag{1}$$

where l and m are the Lame' constants. However, for the cementitious material considered in this study, semi-brittle behavior is assumed and equation (1) is modified to incorporate damage:

$$s_{mn} = 1 f_{mn} f_{kl} \epsilon_{kl} + 2m f_{mk} f_{nl} \epsilon_{kl} \tag{2}$$

In equation (2), $f_{ij}$ are elements of the structural integrity tensor, $f$, which is a measure of material damage. The incremental change in structural integrity in the 'i' principal direction, $df^i$, is given by:

$$df^i = - \left| f_n^i \right|^m k \, d\epsilon_i \tag{3}$$

for $d\epsilon_i > 0$, an incremental tensile strain, and $\epsilon_i > 0$, in a tensile strain region;

$$df^i = 0 \text{ otherwise.} \tag{4}$$

In equation (3), only two calibration factors, k and m, are warranted. The calibration factor k represents fracture propensity while m is a measure of a material's rate of degradation. The total structural integrity of a material at a load is equal to the sum of the incremental structural integrity at that location for the incrementally applied loads.

Unconfined compression tests using the stabilized soil samples were used to determine the elastic material properties, E (elastic modulus), n (Poisson's ratio), ultimate material strength, and calibration factors k and m. Elastic material properties for the 4% stabilized soil sample are shown in Figures 2 and 3. These properties reflect representative values for the thirteen unconfined compression tests conducted on the 4% stabilized soil mixture. The values for the other mixtures are summarized in Table 1.

**Table 1. Monotonic Load Summary**

| STABILIZED SOIL TYPE | # of SAMPLES | Pult (average) | | sult (average) | | Poisson's ratio |
|---|---|---|---|---|---|---|
| | | lb | N | psi | kPa | |
| 4% Cement | 13 | 1721 | 7658 | 183 | 1261 | 0.1 |
| 6% Cement | 11 | 2679 | 11922 | 283 | 1950 | 0.1 |
| 4% Cement + Fibers | 9 | 2906 | 12932 | 295 | 2033 | 0.13 |

TOTAL = 33 SAMPLES

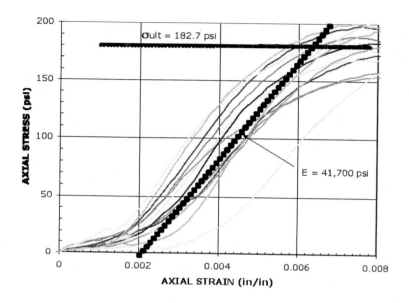

**Figure 2.  Material Elastic Modulus and Ultimate Strength**

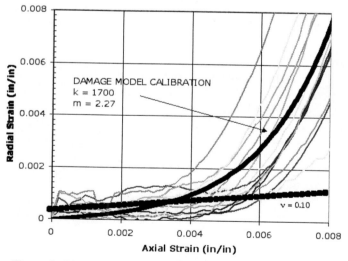

**Figure 3.  Elastic Material Property, n, and Calibration Factors**

To determine the calibration parameters k and m, equation (2) is written in cylindrical coordinates assuming axisymmetric behavior:

$$\begin{Bmatrix} s_{rr} \\ s_{zz} \\ s_{qq} \\ s_{rz} \end{Bmatrix} = \begin{bmatrix} f_{rr}^2(1+m) & f_{rr}f_{zz}1 + f_{rz}^2 2m & f_{rr}f_{qq}1 & f_{rr}f_{rz}(1+2m) \\ f_{rr}f_{zz}1 + f_{rz}^2 2m & f_{zz}^2(1+2m) & f_{qq}f_{zz}1 & f_{zz}f_{rz}(1+2m) \\ f_{rr}f_{qq}1 & f_{zz}f_{qq}1 & f_{qq}^2(1+2m) & f_{rz}f_{qq}1 \\ f_{rr}f_{rz}(1+2m) & f_{zz}f_{rz}(1+2m) & f_{rz}f_{qq}1 & f_{rz}^2(1+m) + f_{rr}f_{zz}m \end{bmatrix} \begin{Bmatrix} \epsilon_{rr} \\ e_{zz} \\ e_{qq} \\ e_{rz} \end{Bmatrix}$$

(5)

The calibration parameters are determined using the analytic $e_{rr}$ - $e_{zz}$ relationship that exists for a cylinder subjected to a uniform compression stress:

$$e_z = -\frac{f_r e_r}{n}$$

(6)

Calibration factors, k and m, associated with $f_r$ are assumed and the corresponding $e_{rr}$ - $e_{zz}$ plot superimposed on Figure 3 using trial and error until a good fit is found (Heymsfield et al. 2007).

## Cyclic Loading Unconfined Compression Tests

Thirty–three unconfined compression tests were conducted on the stabilized soil mixtures to evaluate their material behavior. In addition, forty-eight cyclic loading unconfined compression tests were conducted on the three stabilized soil mixtures, Table 2. The applied load in the cyclic load tests was taken as a percentage of the ultimate material strength determined from the monotonic loading compression tests, Table 1. Loading was applied cyclically until the stabilized soil mixture failed, Figure 1.

**Table 2. Cyclic Load Testing Summary**

| STABILIZED SOIL TYPE | # of SAMPLES | % of Pult |
|---|---|---|
| 4% Cement | 11 | 77 - 88 |
| 6% Cement | 20 | 72 - 92 |
| 4% Cement + 1.5" Fibers | 17 | 69 - 88 |

TOTAL = 48 SAMPLES

The number of cycles to failure as a function of ultimate load percentage is shown in Figure 4. In addition, least square fits are superimposed on Figure 4 to show a tendency that the number of cycles to failure increases as the applied load decreases. However, the low correlation coefficient for each of the three stabilized soil mixtures indicates the high degree of scatter.

**Damage Model Validation**

Results from a cyclic loading test were compared with numerical results using the Valanis damage model. The specific case was a 4% cement stabilized soil specimen subjected to a cyclic load of 141 psi, 77% of the ultimate load, 182.7 psi. The calibration factors found from the monotonic load tests were applied to this situation and numerical results compared with the experimental results, Figure 5. A comparison between the Valanis damage model and experimental results indicates that the damage model captures the general stabilized soil behavior; however that the stabilized soil behavior from the experimental work shows some plastic behavior. To try to capture this plastic behavior within the damage model, residual strains and an increased material stiffness after the initial load cycle were considered. After reviewing cyclic load test behavior, a 0.0018 in/in residual axial strain was assumed after the first cycle. Residual axial strains for each subsequent load cycle were then estimated by assuming an increase of 0.00018 in/in for each following cycle. The corresponding residual radial strain was calculated assuming a linear relationship, the material's initial Poisson's ratio. To account for an increase in material stiffness, a larger material stiffness, E = 115,850 psi, was used for cycles 2 and following. Results of the modified damage model are shown in Figure 5 and shows improved behavior.

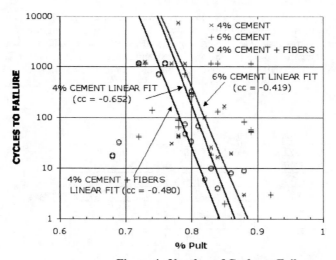

**Figure 4.  Number of Cycles to Failure**

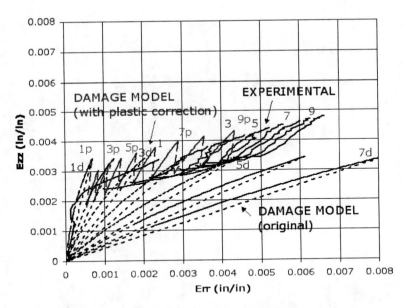

**Figure 5. Damage Model Validation**

**Biaxial Loading**

Although preliminary in nature, the biaxial load problem was considered to demonstrate the damage model's application for this case. Biaxial loading on a stabilized soil layer underlain with an elastic half-space was examined using the Valanis damage model to examine the effects of aircraft braking, Figure 6. The 4% stabilized soil mixture is considered. The layer is 12-in. thick and uses the elastic material properties and calibration factors determined earlier in the article for the 4% stabilized soil mixture. The stabilized soil is underlain by an elastic half-space which has an elastic modulus of 10,800 psi and a Poisson's ratio of 0.35. The original Valanis damage model was used to evaluate this problem and therefore, plastic behavior was not considered. Although the original damage model does not include plastic behavior, the approximate effects of friction can be identified.

To investigate cyclic load behavior, an axisymmetric case was examined in which a 141 psi pressure with 0 friction was applied over a 12-in. radius, Figure 7. This loading represents approximately a C-17 wheel loading. Adjacent soil not included in the finite element mesh was assumed to prevent lateral movement; however allow for vertical displacement. Figure 7 shows damage after load cycle 1 and after 5 load cycles. As the load cycles increases, the extensiveness of material damage, cracking, increases.

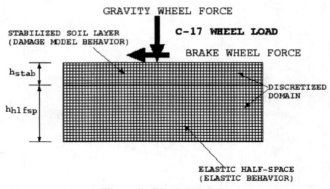

**Figure 6.  Biaxial Loading**

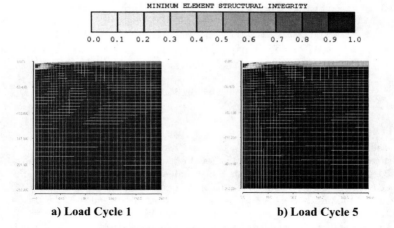

a) Load Cycle 1                          b) Load Cycle 5

**Figure 7.  Damage During Cyclic Loading (Axisymmetric)**

A plane strain analysis was used to examine braking forces induced during aircraft landing. For plane strain, the constitutive relationship in equation (2) becomes:

$$\begin{Bmatrix} \sigma_{xx} \\ \sigma_{yy} \\ \sigma_{xy} \end{Bmatrix} = \begin{bmatrix} \phi_{xx}^2(\lambda+\mu) & \phi_{xx}\phi_{yy}\lambda+\phi_{xy}^2 2\mu & \phi_{xx}\phi_{xy}(\lambda+2\mu) \\ \phi_{xx}\phi_{yy}\lambda+\phi_{xy}^2 2\mu & \phi_{yy}^2(\lambda+2\mu) & \phi_{yy}\phi_{xy}(\lambda+2\mu) \\ \phi_{xx}\phi_{xy}(\lambda+2\mu) & \phi_{yy}\phi_{xy}(\lambda+2\mu) & \phi_{xy}^2(\lambda+\mu)+\phi_{xx}\phi_{yy}\mu \end{bmatrix} \begin{Bmatrix} \varepsilon_{xx} \\ \varepsilon_{yy} \\ \gamma_{xy} \end{Bmatrix} \qquad (7)$$

A pressure of 71 psi, representing a C-17 loading, was applied symmetrically over a 30-in width. Figure 8 shows the extensiveness of material damage considering 30% of the aircraft wheel load as friction and comparing it with 0 braking friction. The damage for the 0 braking case is limited to a region adjacent to the zone directly beneath the wheel load. Conversely, if braking is considered, the region experiencing damage increases.

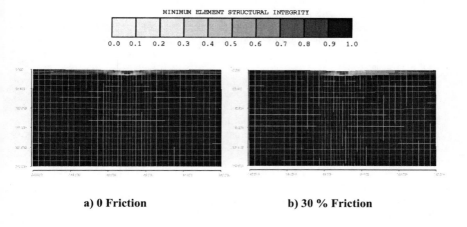

a) 0 Friction                                    b) 30 % Friction

**Figure 8. Damage Considering Friction (Plane Strain)**

**Conclusions**

A total of eighty-one unconfined compression tests were conducted to investigate stabilized soil behavior. The stabilized soil in this study is a mixture of a non-plastic silty sand and a percentage of Type III Portland cement. A damage model was implemented in this study to examine its suitability to model stabilized soil behavior. This study showed that the Valanis damage model may be a viable means to mimic stabilized soil behavior; however, before implementation, the Valanis damage model should be modified to incorporate limited plastic behavior.

Results indicate that damage becomes more extensive as the number of load cycles increases and wheel braking has a significant impact on the region size in which damage develops.

**Acknowledgements**

The research presented in this article was funded by the U.S. Army Corps of Engineers Engineer Research and Development Center (ERDC) and conducted by ERDC and the Department of Civil Engineering, University of Arkansas. The authors are appreciative of these agencies for their support of this research work.

**References**

Harichandran, R.S., G.Y. Baladi, and Yeh, M. (1989). *Development of a Computer Program for Design of Pavement Systems Consisting of Bound and Unbound Materials*; Department of Civil and Environmental Engineering, Michigan State University.

Heymsfield, E. (2006). *User's Manual for STUBBS: A Finite Element Program*, Instruction Report, U.S. Army Engineer Waterways Experiment Station.

Heymsfield, E., Hodo, W.D., and Wahl, R.W. (2007). Development of a Damage Model to Analyze Stabilized Soil Layers Subjected to Repetitive Aircraft Loadings, Transportation Research Board 86th Annual Meeting Compendium of Papers CD-ROM, January 2007.

Huang, Y.H. (2004). *Pavement Analysis and Design*; Prentice Hall, New Jersey.

Peters, J.F. (1997). *Regarding Implementation of a Micromechanically- Base Elastic-plastic Constitutive Model for Soils,* Internal Memorandum, Soil and Rock Mechanics Division, Geotechnical Laboratory, USAE Waterways Experiment Station, Vicksburg, MS, 1997

Peters, J.F., Wahl, R.E., & Meade, R.B. (1997). *User's Manual for STUBBS: A Finite Element Program for Geotechnical Analysis*, U.S. Army Engineer Waterways Experiment Station, GL-97-1.

Valanis, K.C. (1988). *A Theory of Damage of Brittle and Semi-Brittle Materials*, AFOSR Bolling Air Force Base, ENDIC-005-AFOSR-1988.

Unified Facilities Criteria (2001). UFC *Pavement Design for Airfields*, UFC 3-260-02, 30 June 2001, Washington, DC 20314-1000.

Raad, L, and Figueroa, J.L. (1980). Load Response of Transportation Support Systems," *Transportation Engineering Journal*, ASCE, Vol. 106, No. TE1, pp. 111-128, 1980.

# A Parametric Sensitivity Analysis of Soft Ground Arrestor Systems

E. Heymsfield,[1] W.M. Hale,[2] and T.L Halsey[3]

[1]University of Arkansas, Department of Civil Engineering, 4190 Bell Engineering Center, Fayetteville, AR 72701; PH (479) 575-7586; FAX (479) 575-7168; email: ernie@uark.edu
[2]University of Arkansas, Department of Civil Engineering, 4190 Bell Engineering Center, Fayetteville, AR 72701; PH (479) 575-6348; FAX (479) 575-7168; email: micah@uark.edu
[3]University of Arkansas, Department of Civil Engineering, 4190 Bell Engineering Center, Fayetteville, AR 72701; PH (479) 575-2215; FAX (479) 575-7168; email: thalsey@uark.edu

## Abstract

The Federal Aviation Administration (FAA) requires airfields to have a 1000-ft. runway safety area beyond the design runway length for aircraft overruns. However, at many locations, this requirement cannot be satisfied because of natural or man-made barriers. An alternative solution to the 1000-ft. runway extension is to use an engineered material arresting system (EMAS). An EMAS system is designed to significantly reduce an aircraft's stopping distance during an aircraft overrun without significant passenger discomfort or aircraft damage. The FAA computer code ARRESTOR is used in this study to evaluate aircraft stopping distance as a function of characteristic parameters that are typically considered in an EMAS design. This study considers the sensitivity of stopping distance as a function of arrestor material compressive strength behavior, aircraft characteristics, and aircraft type for a given arrestor bed geometry. During this study, 121 computer overrun simulations are performed to develop a sensitivity study of the EMAS. The sensitivity of aircraft stopping distance is examined as a function of a specific study parameter (arrestor material characteristic or aircraft characteristic). Results are normalized using typical design parameters as a basis. Results for stopping distance are summarized as plots as a function of variability in material strength. The sensitivity analysis conducted in this work identifies critical EMAS design parameters and therefore, provides a means to optimize an EMAS design.

## Introduction

The Federal Aviation Administration (FAA) requires airfields to have a 1000-ft. runway safety area beyond the design runway length for aircraft overruns. At many locations, this requirement cannot be satisfied because of natural or man-made barriers. At these locations, an alternative solution is to use an engineered material arresting system (EMAS). An EMAS is designed with the intent to significantly reduce an aircraft's stopping distance during an overrun.

The FAA requires that an EMAS design be validated using a design method which can predict the arrestor material's performance (FAA, 1998). Instead of conducting an extensive full-scale overrun test program on the modified system, a numerical approach using the FAA computer code, ARRESTOR (Cook et al, 1995) can be used. ARRESTOR is an enhanced version of the computer program initially developed for the U.S. Air Force (Cook, 1985). Consequently, multiple overrun simulations can be easily performed as a function of EMAS design characteristics and aircraft type to optimize an EMAS design.

**Background**

An overrun occurs when a plane stops beyond the runway length either during landing or an aborted takeoff. Overruns may be due to inadequate takeoff power, weather conditions, incorrectly set flaps, or pilot error. The National Transportation Safety Board and the International Civil Aviation Organization recorded 26 overruns between 1975 and 1987 (White et al, 93). The majority of aircraft (90%) involved in overruns have an exit speed of 70 knots or less and stop within 1000 feet of the runway end (FAA, 98). Overruns can result in loss of life and severe aircraft damage. In response, the Federal Aviation Agency now requires a 1000-ft safety area beyond the runway length to stop an overrunning aircraft. However, many airfields are prevented from including a 1000-ft safety area because of natural or man-made barriers. In 2002, there were approximately 350 runways with inadequate runway safety areas (San Filippo and DeLong, 2002).

The FAA allows airfields that have insufficient runway safety areas to use an engineered materials arresting system (EMAS) instead of restricting aircraft. FAA requirements for an EMAS are described in the FAA Advisory Circular AC 150/5220-221 (FAA, 98). An EMAS is designed as a passive system constructed using a material that crushes under aircraft weight and that absorbs energy from the aircraft through drag forces to bring an aircraft to a stop. The material normally has a weak compressive strength (less than 30 psi), however needs to be able to support emergency vehicles with limited material deformation. Besides crushable and energy absorbent, the material needs to be durable. Natural materials such as sand, clay, and gravel have been investigated. However, sand and clay materials are dependent on weather conditions and therefore are not considered as a viable solution. Although gravel is highly energy absorbent, it is loose and can be ingested into aircraft jet engines. Instead of natural materials, engineered materials can be made to be highly energy absorbent and weather resistant. Early engineered materials for EMAS included plastic material types and foamcrete. In the 1990's, the Engineered Arresting Systems Corporation (ESCO) developed an EMAS using closed-cell foamed concrete (San Filippo and DeLong, 2002). The cost of constructing an EMAS with closed-cell foamed concrete varies however is dependent on the runway's current safety area condition and aircraft usage. Total construction costs for an EMAS varies. In the U.S., an EMAS can be estimated at between $2,000,000 and $5,000,000 (www.esco-usa.com).

## Arrestor Code Description

The ARRESTOR computer code evolved from the FITER1 computer code developed in 1983 (Cook, 85). FITER1 was formulated to investigate landing gear loads and aircraft deceleration of US Air Force fighter landings on compliant surfaces. The compliant surface, soft ground, in FITER1 was restricted to sand or clay. In addition, aircraft characteristics considered in the code were limited. In a 1987 FAA study on developing an arresting system at airfields, FITER1 was used as a basis and then revised to include foam as an arrestor material (Cook, 1987). The new code, ARRESTOR, was validated using full-scale tests including an instrumented Boeing 727 aircraft entering a phenolic foam bed (White et al, 1993). In later studies, the Engineered Arresting Systems Corporation (ESCO) made further revisions to the ARRESTOR computer code to model a specific EMAS product constructed using foamed concrete blocks (San Filippo and DeLong, 2002).

## Arrestor Code Formulation

The ARRESTOR code uses a mathematical model based on force equilibrium on the aircraft strut in the vertical and horizontal directions. The strut model is shown in Figure 1. The base of the fully compressed strut defines a reference line and the fully extended strut represents the strut at zero load. The compressed strut represents actual strut extension while the aircraft traverses the EMAS.

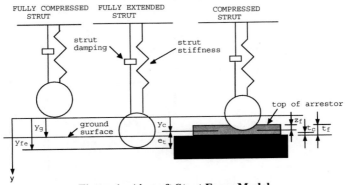

**Figure 1. Aircraft Strut Force Model**

where
$y_g$ =     y coordinate of ground surface
$y_{fe}$ =    y coordinate of fully extended strut
$y_c$ =     y coordinate of compressed strut
$z_f$ =     arrestor penetration
$t_f$ =     thickness of arrestor material
$t_c$ =     thickness of crushed arrestor material
$e_t$ =     strut contraction

From Figure 1, arrestor penetration is equal to:

$$y_c = y_{fe} - e_t, \text{ and } z_f = y_c - (y_g - t_f),$$  (1)

and the maximum arrestor material penetration is:

$$z_f(\text{maximum}) = t_f - t_c$$  (2)

If $z_f$ is greater than 0, the strut has penetrated the top of the arrestor and forces are induced on the aircraft strut. The vertical force on the aircraft is a function of the spring contraction and rate of contraction from the strut's fully extended state:

$$P_{strut} = f(e_t, \dot{e}_t)$$  (3)

If the force in the strut, $P_{strut}$, exceeds the full-thickness arrestor compression strength, $P_{arrestor}$, the arrestor material will begin to crush, Figure 2.

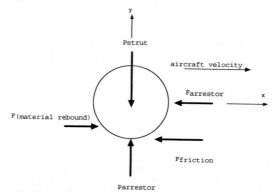

**Figure 2. Free-Body Diagram of Aircraft Wheel**

Figure 2, shows the aircraft traveling in the +x direction. For this directional movement, forces acting in the negative x direction work to decelerate the aircraft by dissipating energy. These energy dissipation forces include $F_{arrestor}$, the drag force induced by the arrestor material, and $F_{friction}$, the frictional force at the wheel-arrestor interface. Conversely, the horizontal force, $F_{rebound}$, acts in the positive x direction. $F_{rebound}$ develops when the arrestor material rebounds after the aircraft wheel load traverses a point. Aircraft deceleration is calculated using Newton's Second Law:

$$m\ddot{x} = F_{(materialrebound)} - F_{friction} - F_{arrestor}$$  (4)

## Drag Force Calculation

Drag forces, which produce aircraft deceleration, occur on planes perpendicular to the aircraft wheel velocity, y', shown in Figure 3. When there is no wheel slippage, y' is collinear with y. Consequently, drag forces develop along the component of the wheel width, w, and wheel length, L, in the y' plane, Figure 3.

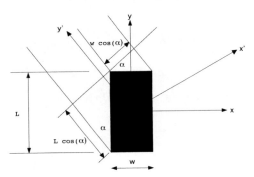

**Figure 3. Aircraft Wheel Plan View**

In Figure 3, a defines the wheel velocity direction. For a „ 0, there is wheel slippage. When there is wheel penetration into the arrestor material, the arrestor material induces a drag force on the y' wheel face and a side force on the x' face of the wheel. Deceleration is assumed to result from the compressive strength along the vertical plane of the arrestor material. A side force is assumed to develop only during wheel slippage. Side friction is assumed negligible and deceleration forces occur only on the y' plane perpendicular to the wheel velocity. The drag force and side force are proportional to the areas in which they act.

The drag force acts on $A_{y'}$:
$$A_{y'} = [w * \cos(a)] * z_f \tag{5}$$

and the side forces on $A_{x'}$:
$$A_{x'} = q * R^2 = R^2 \tan^{-1}(\frac{L}{2(R - z_f)}) \tag{6}$$

where L is the wheel length at the wheel-arrestor interface, Figure 4:
$$L = \sqrt{8R * z_f - 4 * z_f^2} \tag{7}$$

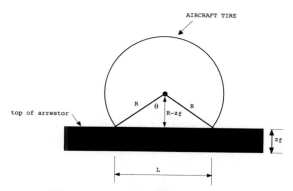

**Figure 4. Aircraft Wheel Elevation View**

## ARRESTOR Code Analysis

The FITER1 code was revised for the FAA by Cook et al to evaluate the performance of soft ground arrestor systems (Cook et al, 95). The ARRESTOR code assumes a foam material for the arrestor material and is limited to three aircraft types: B707, B727, and B747. An overview of an overrun analysis using ARRESTOR is shown in Figure 5. The program is organized so that the user is queried through the ARRESTOR Main Menu. The menu includes three input submenus: "Define Material", "Design Arrestor", and "Select Aircraft". Each of these options includes a screen for the user to input required information. Aircraft Data, which is a function of the specific aircraft, is fixed and is a function of the aircraft specified through the "Select Aircraft" menu. After data input, the user initiates an arrestor analysis using the ARRESTOR "Start Calculation" sub-menu. Output data are displayed as plots using the ARRESTOR "Result" Menu.

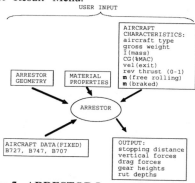

**Figure 5. ARRESTOR Input – Output Overview**

## Sensitivity Analysis

A sensitivity analysis was conducted investigating aircraft stopping distance as a function of arrestor material and aircraft characteristics. Although, arrestor bed geometry influences aircraft stopping distance, a single geometry was considered for this study. The typical EMAS configuration includes an asphalt ramp after the runway threshold, an inclined arrestor material section, and in some applications a second bed using a stiffer material. The arrestor bed geometry considered in this study is shown in Figure 6. Aircraft stopping distance is determined in this study as a function of arrestor material compressive strength and aircraft characteristics: gross weight, mass moment of inertia about the pitch axis, center of gravity as a percentage of mean aerodynamic chord, runway exit speed, and total reverse thrust. The analysis investigates two aircrafts, B727 and B747, along with two arrestor material types.

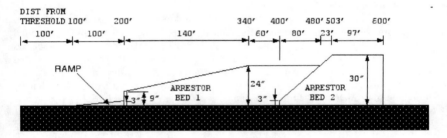

**Figure 6. Arrestor Bed Geometry**

## Arrestor System Material

Two arrestor bed materials were considered, a phenolic foam and a low density concrete foam. For the arrestor bed in Figure 6, a sensitivity analysis was conducted considering Bed 1 and Bed 2 using the same phenolic foam material. Phenolic foam is commercially available and manufactured in 48" x 96" x 3" panels. Varying arrestor bed depth is created by stacking the phenolic foam panels. A second case was also studied in which the beds were created using low-density concrete foam. For this second case, a stiffer material was used for Bed 2 than the material in Bed 1. Compression strengths of the materials are shown in Figure 7. Figure 7 shows the nominal strength of each of the materials along with a 20% variance from the nominal value. Both materials have similar foam material behavior where the load significantly increases when the test specimen becomes fully compressed. The sensitivity analysis was conducted using ±20% values as lower and upper bounds for these materials.

## Sensitivity Analysis

Two sensitivity studies were conducted. In the first study, aircraft parameters were varied by 20% and stopping distance determined using upper and lower bound arrestor material strengths. In the second study, nominal values were used for the aircraft parameters and material strength was varied.  Two aircraft types, B727 and B747, were considered in this sensitivity analysis.

Aircraft characteristics for each aircraft type were varied to study the significance of a specific aircraft parameter to stopping distance using twenty-one overrun simulations.  These aircraft characteristics include; gross weight, mass moment of inertia about the aircraft pitch axis, center of gravity as a percentage of the mean aerodynamic chord, fraction of total reverse thrust, and main gear wheel friction.  These parameters were varied by decreasing their nominal values by 20%, to represent a lower bound, and except for reverse thrust, increasing their nominal values by 20% to represent an upper bound.  The results of the study are shown in Figure 8.  The nominal values for each specific parameter are included in parentheses adjacent to the aircraft parameter description.  The study of the five aircraft parameters shows that stopping distance is most dependent on gross aircraft weight. In this study, a variation of ±20% of the gross aircraft weight changes the stopping distance by approximately 15% .

A second sensitivity analysis was conducted using 100 overrun simulations to investigate stopping distance as a function of arrestor material strength.  Nominal values were used for the two aircraft types, B727 and B747; however, material compression strength, shown in Figure 7, was varied.  Figure 9 implies that decreasing the arrestor material compression strength results in increasing the stopping distance.  In addition, increasing the compressive strength of the low-density concrete foam decreases stopping distance.

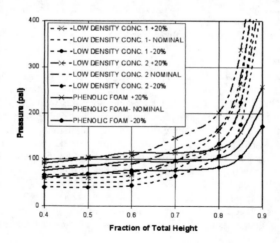

**Figure 7.  Material Compression Strength**

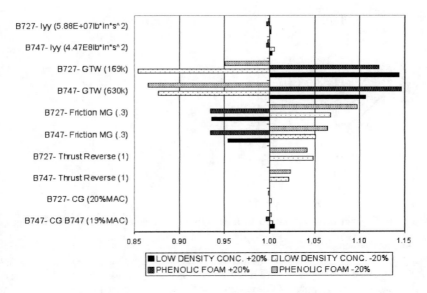

**Figure 8. Sensitivity Analysis as a Function of Aircraft Characteristics**

## Conclusions

Engineered material arrestor systems (EMAS) are a viable alternative at airports that have inadequate runway safety areas for aircraft overruns. These systems are designed to stop an aircraft by creating drag forces on the aircraft's nose and main gear. Typically, phenolic foam or low-density concrete materials are used in arrestor beds because of their high energy absorption and cohesiveness. The computer code ARRESTOR was used in this study to investigate critical parameters in the design of a soft ground arrestor system considering a specific bed geometry. Twenty-one overrun simulations were conducted to investigate the sensitivity of stopping distance to five aircraft parameters (pitch mass moment of inertia, weight, friction, reverse thrust, center of gravity location). Of the five aircraft parameters, aircraft weight was found to have the greatest impact on stopping distance. In a second study, 100 computer overrun simulations were used to analyze stopping distance as a function of arrestor material compressive strength. Reducing the material compressive strength increased the aircraft stopping distance. Although this study was conducted considering a specific soft ground arrestor configuration, the approach can easily be applied to multiple configurations to develop an optimal system as a function of cost. Currently the ARRESTOR code is limited in the number of aircraft types that can be analyzed; however, in future work, the authors hope to be able to analyze stopping distance as a function of a larger aircraft fleet.

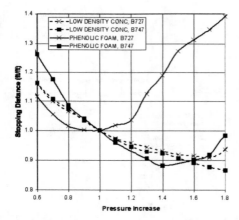

**Figure 9. Sensitivity Analysis as a Function of Material Compression Strength**

### Acknowledgements

The research presented in this article was funded by the Strong Company and the Mack-Blackwell Transportation Center. The authors are appreciative of these agencies for their support of this research work.

### References

Cook, R.F. (1985).*Aircraft Operation on Soil Prediction Techniques.* Technical Report ESL-84-04, Vol. 1 and 2, U.S. Air Force Engineering and Services Center, Tyndall Air Force Base.

Cook, R.F. (1987). *Soft Ground Arresting Systems*, FAA Technical Report: DOT/FAA/PM-87/27, Department of Transportation/Federal Aviation Administration

Cook, R.F. (1995). Teubert, C.A., and Hayhoe, G., *Soft Ground Arrestor Design Program*, Technical Report # DOT/FAA/CT-95, National Technical Information Service, Springfield, VA

Federal Aviation Administration, (1998).*Engineered Materials Arresting System (EMAS) for Aircraft Overruns*, Advisory Circular 150/5220-22.

San Filippo, W.K. and DeLong, H., (2002), "Engineered Materials Arresting System (EMAS): An Alternative Solution to Runway Overruns," *Air Transport*, pp 257-265.

White, J.C., Agrawal, S.K., and Cook, R.E., (1993). *Soft Ground Arresting System for Airports*, Federal Aviation Administration, DOT/FAA/CT-93/80.

# Worldwide Standards for Concrete Mixtures for Safe and Durable Runways and Taxiways

James M. Shilstone, Sr. P.E., Concrete Engineer[1]

[1]Shilstone & Associates, Glen Lakes Tower #105, 9400 N. Central Expressway, Dallas, TX 75231; PH (214)361-9681; FAX (214)361-361-7925; e-mail: jim.shilstone@shilstone.com

## Abstract

The technology that led to optimized concrete paving mixtures was born of a need to develop a method to specify and control the quality of concrete mixtures for construction of a $600 million office building in Riyadh, Saudi Arabia in the early 1970s. However, there were no materials standards in the kingdom. It was necessary for this author to learn how to achieve that goal using the non standard, locally available materials. The research was done in a laboratory in Athens, Greece. The results included the introduction of a nomograph on which six independent variables can be plotted and used to assess concrete mix performance. That technology is being applied today for construction of buildings, highways, and airfields all over the world. The first verification of the technology for airfield construction was done at the Jeddah International Airport in Saudi Arabia.

## Introduction

Concrete is an international material. The stone and sand on the earth's crust differ but only within quantifiable limits. Cementitious materials vary minimally and facilitate strength development to meet engineering and long-term performance requirements. International producers of chemical admixtures apply similar technologies to support local cementious materials. Potable water does not vary to a significant degree. When properly selected and used, locally available concrete-making materials can meet a required strength and be durable for 50 or more years. High quality, durable concrete can be produced anywhere in the world by paying close attention to the gradation of the aggregate matrix. This paper describes the development and application of several graphical tools to illustrate the gradation and develop a workable, quality concrete mixture.

## Origin of the Technology

In late 1960s, the king of Saudi Arabia recognized the need to meet future challenges to be active in the world of nations. Since there was no Saudi Arabian construction industry or leadership to manage the planned massive construction program, the United States government was asked to undertake the construction management task. That mission was assigned to the U. S. Army Corps of Engineers. US architects and engineers were selected to design the projects and aid in construction management. One project was the $660 million Headquarters for the Saudi Arabian National Guard (SANG) in Riyadh. Architect / Engineer Leo Daly of Omaha, NE was selected to design and supervise construction. The only reasonable architectural solution was concrete with architectural quality finishes. Daly retained Architectural Concrete Consultants, Inc. (ACC) of Houston to develop a solution. This author headed that company. ACC went to Riyadh to evaluate the available materials – especially the aggregates - and develop a program to make best use of those materials to achieve the architectural intent.

Work in Saudi Arabia was not new to ACC as they helped develop the solution to construction of the University of Petroleum and Minerals in Dhahran several years earlier. As was found in Dhahran, there were no materials standards in the kingdom. The US aggregate grading standards were not appropriate since they did not take into consideration the impact of local climactic conditions and environment. ACC found usable coarse aggregate and pea gravel at one site and fine sand at another. Samples were collected and flown by the US Air Force to Athens, Greece. The staff of a former Corps of Engineers laboratory aided in the research to develop the new technology for proportioning aggregates for the cast-in-place, architectural concrete mixture.

Aggregate gradations and specific gravities were determined. The coarse aggregate and pea gravel were blended to facilitate use in a 2-bin concrete batch plant. Different blends of the coarse size and sand were cast and evaluated for workability, finishability, and strength. No admixtures or air entraining agents were used. Care was taken to determine the water requirement for each batch. It soon became apparent that the optimum mix had the following characteristics: it required the least amount of water, responded best to vibration, and produced the highest strength.

ACC reviewed earlier research. The work done by Weymouth (Weymouth 1933) and Powers (Powers 1968) identifying three aggregate sizes shown in Figure 1 made it possible to see a relationship between the three segments – coarse, intermediate, and fine - of the combined aggregate.

One mix was the optimum and the basis for further study. From this, ACC recognized that the amount of fine aggregate required for a mixture was a function of the relationship between the coarse and intermediate aggregate. That concept was the foundation of the Coarseness Factor Chart (CFC) nomograph. Seven independent variables (up to four aggregates and three cementitious materials) can be coordinated and reported as a single point on the Chart. Figure 2 is the first published results of the Athens research Coarseness Factor Chart (Shilstone, 1990). Studies of other mixes with known performance are identified by dots on the chart. The x-axis is the Coarseness Factor, the percent of aggregate retained on the No. 8 sieve that is also retained on the 3/8-inch sieve. The y-axis is the percent minus No. 8 sieve. The term "Workability" was applied because it was the constituent that provided mobility for a mixture. There is an adjustment made on the fine aggregate based upon the cementitious materials content. The original mix was cast with 6.0 US bags of cement (564 lbs.). A difference of one US bag of cement is approximately 2.5 percentage points on the Workability chart. If a mix contains 5 US bags (470 lbs) cement, the Workability is decreased 2.5 points.

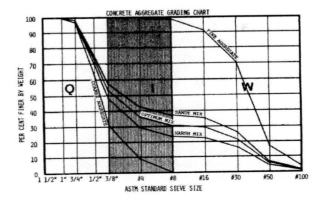

**Figure 1. Aggregate Grading and Blend (Shilstone 1990)**

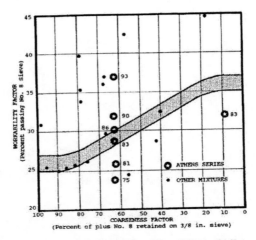

**Figure 2. SANG Initial Chart and Other Mixes (Shilstone 1990)**

The diagonal bar in Figure 2 is known as the "Trend Bar" that divides the sandy mixtures and the very coarse mixtures. The Trend Bar location reflects the changes in fine aggregate requirement as the two coarser sizes become finer. A "100" mixture has no intermediate particles so it is gap-graded. A "0" mixture is a pea gravel mix with nothing retained on the 3/8-ich sieve. A Bar is used to reflect the effects of aggregate shape and texture on mix workability. The dots in the Trend Bar, or near it, represent intentionally gap-graded mixtures – often using masonry sand – to assure a high density of coarse aggregate when the concrete was sand blasted. The SANG mixes were to be used in that manner.

The final SANG report was issued in September 1975 (Shilstone 1975). The third section was titled "Conclusions" with the prophetic statement:

> *"In summary, this limited scope and time study has been very rewarding and caused, through failure to comply with expected results, an in-depth research of the state-of-the-art and the evolution of a new concept which bares potential of being a significant breakthrough in simplicity for concrete mix designed technology in all structural work using hard rock aggregates appropriate for building construction."*

### First Airfield Application

A few months after the technology was developed, the Hochtief Project Manager who was leading construction of the new Jeddah International Airport contacted ACC. The runways and taxiways were designed by a US engineering firm with the work to be done using US specifications such as ASTM C-33 for the aggregate quality and grading. The slip-formed paving in that era involved tongue and groove centerline joints. The paving machine is shown in Figure 3. The segregation of the coarse aggregate can be seen to the right of the mix distribution screw. The tongue edge joints tended to sag or fall off of the slab. Wooden strips with vertical supports were placed under the extruded edge to serve as "crutches" - a costly solution (see Figure 4).

**Figure 3. The Paving Machine (Shilstone)**

**Figure 4. Tongue with "Crutches" (Shilstone)**

A long section of runway (about 14" thick) was being removed because the concrete mix was not sufficiently plastic to be placed without honeycomb using the paving machine. Upon visiting the batch plant with the Project Manager, the problem became obvious. After the contractor screened the bone-dry local sand and gravel, the coarse and fine aggregate was placed in stockpiles for later use. The US specification was not appropriate, as it did not recognize how dune sands are moved by the wind, leading to wind-blown segregation and its effect on the concrete sand stockpile. It was also found that small "nodules" were formed by a coalescence of clay particles. They could be identified when sand grading was determined after washing. These particles contributed to high water demand and rapid slump loss. The concrete mixtures had to be in place within 20 minutes after adding water to the mix.

ACC explained the new technology to site engineers and recommended that both concrete aggregate piles be reprocessed and separated into three sizes based upon what was learned during the SANG research (Shilstone, 1990):

- The coarse aggregate was that aggregate retained on the 3/8" sieve,
- The intermediate aggregate that passed the 3/8" sieve and was retained on the No. 8 sieve,
- The fine aggregate stockpile consisted of all aggregate passing the No. 8 sieve.

The SANG research describes the relationship between the three aggregate sizes. The coarse aggregate fills the major voids in the mixture. The intermediate aggregate fills in the major voids between the coarse aggregate and the fine aggregate completes the process for the aggregates that are then coated with a mixture paste. With an understanding of the requirements for each size, the Courses Factor Chart (CFC) developed during the research was explained. Finally, the methodology for obtaining the optimum concrete mixture and its relationship to the CFC was discussed.

Approximately a year later, ACC visited the project site and met with the Project Manager who said the system that had been explained to him was used and produced the desired results. The problems had been resolved and there was no need to place "crutches" under the tongues.

## Expansion of the Technology

A major lesson was learned from the Jeddah experience that can be correlated with slip-formed concrete paving cast today. Before starting construction, the characteristics of the locally available materials must be considered. They must meet the quality requirements of an agency or other standard, and when combined, the aggregate grading must meet an established combined aggregate grading (CAG). Though aggregate is traditionally divided into stockpiles at the production quarry or pit for concrete, asphalt or other application, they are all one part of the basic concrete-making material. If the "asphalt" aggregate meets quality standards for concrete and fits the CAG requirements, it may be used unless it is soft and will polish where it can affect friction.

Over the past 35 years, the technology has been expanded. A very important part of that research was the study of old documents. Many of those documents support this new technology. They also contribute to a better modern understanding of concrete mixtures. Following are summaries of some of those documents that contributed to concrete pavements and other structures that have performed well in severe environments for 50 or more years.

### 2<sup>nd</sup> Edition, Design and Control of Concrete Mixtures (DCCM 1927)

*2nd Edition, Design and Control of Concrete Mixtures (DCCM 1927)*

This was issued in January 1927, following completion of the 10-year Lewis Institute-PCA research. It provides an excellent foundation for understanding the practical implications of that research. The last page provides a list of 18 projects completed in that era, with technical data about the construction and the mixture designs. Some of those projects are still in service and can be evaluated. This document is especially important to help a reader recognize the relationship of water/cement ratio and strength. The other important point that is made is the correlation of durability and impermeability. The introduction includes the following statement:

> "The fundamental requirements of practically all concrete are strength, durability and economy. These can be obtained only by the proper selection of the materials, and intelligent design of the mixture and the adoption of proper methods of mixing and placing the concrete and protection during the curing period."

The next section is titled **"Water--Cement Ratio Strength Law."** The more significant statement is:

> "For given materials and conditions of manipulation, the strength of concrete is determined solely by the ratio of the volume of mixing water to the volume of cement so long as the mixture is plastic and workable."

The third section title is **"Effect of Water-Cement Ratio on Other Qualities."** The opening statement recognizes that the water-cement ratio correlates with other engineering qualities including flexural strength, resistance to wear, modulus of elasticity, and bond between the concrete and reinforcement:

*"The resistance of concrete to the severe conditions of weathering and to the action of sulfate waters is determined largely by the degree of impermeability. Impermeable concrete requires non-porous aggregates thoroughly incorporated in a cement paste that itself is impermeable. An impermeable paste requires in turn a low water-cement ratio; the thorough incorporation of aggregates necessitates a plastic, puddleable mix. Thus proper control of the mixing water is the vital factor in the production of durable concrete."*

Early ideas of "low water-cement ratio" were very different from those considered "low" today. Table I in the 2$^{nd}$ Edition describes mixtures to be used for *"roadways, piles, pressure pipe and tanks; thin structural members in severe exposure." Walls, dams, piers, etc., where exposure to severe action of water and frost."* It was suggested that construction of such projects required 3000 psi concrete at 28-days with a maximum mixing water content of 6.0 gal per sack of cement. Using the volumetric water-cement ratio of 6.0 gal volume of water to one cubic foot of loose cement volume produced a w/c of 0.80. Today with mass units, 0.53 would be appropriate. A similar water-cement ratio was found in structures cast in that era and still in service.

## *Weymouth / Powers*

The work by C. A. G. Weymouth (Weymouth, 1933) and the follow-up by Treval Powers in his text ***Properties of Fresh Concrete*** (Powers, 1968) describe the effect of gaps in the grading of the combined aggregate on mix segregation. This was the foundation for the Coarseness Factor Chart because it described the relationship between the coarse and intermediate sizes. Weymouth identified one of the major problems in concrete construction – segregation. T.C. Powers clarified Weymouth's work with the following explanation:

*"Although Weymouth was originally interested in concrete mixtures, and was concerned with the effect of particle interference on water requirement and workability, including the tendency of different-sized particles to segregate during handling, he illustrated his concept in terms of dry mixtures of aggregates, using the model shown in Fig [5]. Fig. [5]"a" represents, in two dimensions, a mixture of two sizes of particles. The larger particles are few and are widely separated by the smaller particles; the average clear distance between is considerably greater than the diameter of the smaller particles. . . In diagram "b", the relative number of the larger particles is greater, and the average distance between them is supposed to be just equal to the diameter of the smaller particles. According to Weymouth, for the composition represented by either diagram, the mixture can be stirred without changing the uniformity of the "void pockets" defined by the smaller particles.*

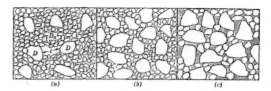

**Figure 5. Weymouth's Examples (Powers 1968)**

*"In Fig. [5] "c", the concentration of the larger particles is such that the average clearance between them is less than the diameter of the smaller particles, making it impossible for the interstitial spaces of the larger particles to be filled uniformly with the smaller. Weymouth said that when such a one-layer mixture on a tray is stirred, there is a tendency for the two sizes to run into separate groups, each of its own kind; in other words, stirring such a mixture tends to produce segregation of the two sizes. To apply this observation to a deep mass, he visualized a given size group as forming a sort of grid structure through which the smaller particles move both horizontally and vertically during manipulation of the mixture; so long as they can move freely, the mass remains homogeneous, but if the larger particles interfere with the movement of the smaller, segregation occurs, and large void pockets are developed with a great loss of strength and workability."*

The higher specific gravity coarse aggregate can settle to the bottom of a concrete pavement during the placement and consolidation process. This condition was found when over sanded mixtures that were not balanced segregated. Shortly after project completion cracks developed over what had been the vibrator trails. When consolidated, an optimized concrete mix will take a thixotropic set. This is a physical set due to a physical interlock of the well-distributed aggregate and an adequate amount of paste. The "tongues" sagged in the Jeddah mixes because that type of set did not occur. Today we recognize the segregation problem during placement and/or as edge slump or slough.

### *Texas Highway Department, Information Exchange, Issue 58 (1938)*

This report describes 3 years of laboratory research to better understand the relationships between the aggregate particles and portland cement. The purpose was to determine how the aggregate should be combined for the pavement mixtures and how much cement would be required. Originally paving mixtures required a specified amount of cement. Results of tests indicated the strengths were much higher than needed.

The gradings and other characteristics of the individual fine and coarse aggregates were studied. That was followed by a study of the grading of the combined aggregate. There was special study of the voids and gaps in the CAG of a mixture. They cited the fact that a change in fine aggregates from 39% to 35% made a major difference in water requirement for a given consistency. Thus there was a reduction in portland cement. The research followed the principles established by

Professor Abrams and of the Lewis Institute (Abrams, 1925). They were assisted by staff from the Bureau of Public Roads. Figure 6 (Information Exchange, 1938) below is a graphical description of the need for a well-graded combined aggregate reported in Texas in 1938.

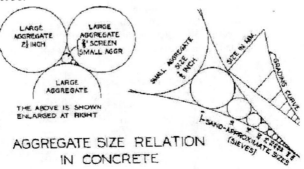

**Figure 6. Aggregate Grading per Texas Highway Department**

Fourteen highways were constructed to apply what was learned in the laboratory. The motive was to improve the product and, at the same time, reduce cost. Following the research and highway construction, the State Auditor reported the actual cost savings was $1 million over two years due to a general reduction in the cement factor.

The 14 highways studied in the report were constructed between 1932 and 1934. They used cement factors varying from two to four US bags of cement per cubic yard. Data from each project was provided in the report along with graphics showing the change in strength between cores cast at the completion of the project and cores taken six years later. A brief description of each project was provided with material sources, batch size, cement factor, water factor in gallons per bag, cylinder strength at 28 days, beam strengths, and the average strength and number of cores. The project completion and 6-year core strength are shown in Table 1.

The data in `Table 1 covers six projects with a total length of 51-miles using 4.0 bags (376 pounds) of cement per cubic yard. Based upon that information, the average 28-day compressive strength per pound of cement was 11.81 psi per pound. At 6-year age, the strength of the cores is 16.39 psi per pound of cement. The cement from that era is often referred to as "old cement."

### Changes in Cement Characteristics

The characteristics of "old cement" were changed in the early 1970s to help concrete contractors accelerate their construction process and better compete with steel contractors. The C3S, C2S, and fineness of grind were the major factors changed. The C3S (Alite) accelerates early strength gain for form stripping and C2S (Belite) provides the component to increase strength over the long term. The finer grind helped expose more Alite to water and increase early strength gain. That change resulted in an increased early strength and a reduction in long-term strength. Recent comparisons with varying cement contents and air-entrainment provided compressive strengths at 28 day age: 423 lbs/cuyd and no entrained air = 6200 psi and 545 lbs/cuyd and 5% air = 6900 psi. The efficiencies were 14.59 psi per pound and 12.66 psi per pound.

There appears to be a general correlation between the old cement and new cement when the potential air-entrainment of the old cement is considered. As will be noted in the following section on air-entrainment, the old cements were found to produce 3 to 4% entrained air in concrete due to the presence of certain grinding aids used at that time.

**Table 1. 1938 Texas Highway Department Field Test Results**

| Number of cylinders | 1,031 |
|---|---|
| Average 28-day strength (psi) | 4,440 |
| Number of beams | 764 |
| Average 7-day strength (psi) | 601 |
| Average 28-day strength (psi) | 764 |
| Number of 6-year cores | 337 |
| Average strength | 4,487 |
| Average strength 6-years (psi) | 6,163 |

### The Shilstone Companies' Software

In 1988, an MS DOS based software program named **seeMIX** was developed. It is now in its third generation. This software provided graphical analysis methods using the CFC and added the asphalt industry's modified 0.45 Power curve and the Percent of Aggregate Retained on Each Sieve. It has been found that each chart has special applications.

- The CFC shows over-all relationships,

- The 0.45 Power reflects trends and sharp changes, and

- The Percent of Aggregate shows details.

Experience has indicated that the CFC is the most important element but information from the other two graphs and their supporting data can over-rule what might appear to be a good mix. This is especially true for the Percent of Aggregate information. The sum of the percent of aggregate retained on two adjacent sieves should not be less than 13% of the combined aggregate. It has been found that when this occurs, that mix will segregate significantly. Figure 7 includes 5 "Zones." Each indicates potential mix performance as follows:

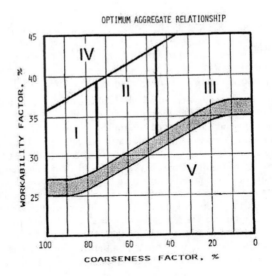

Figure 7. Coarseness Factor Chart (CFC)

**Zone I** mixes with a Coarseness Factor of 75 or more will tend to segregate, experience edge slump or slough, and/or spall or scale.

**Zone II** mixes with Coarseness Factor between 45 and 75 will generally perform well. Caution should be exercised if a mix plot falls near Zones I and IV.

**Zone III** mixes with Coarseness Factors less than 45 are similar to Zone II but for smaller top size aggregate.

**Zone IV** is the triangular portion of the chart above Zones I and II. Such mixes will generally exhibit the most undesirable features of concrete including cracking, spalling, and scaling.

**Zone V** is for mixes below the Trend Bar. These are too rocky to be readily placed and consolidated.

The CFC has been widely tested in the field. The value of the chart became clear during construction of a large project that required two ready mixed concrete producers to supply the needs. Mixture proportions were selected by the responsible testing laboratory based upon statistical data. When the effects of those mixes were evaluated using the CFC, it was found that the aggregate gradings from the two sources differed widely. One supplier's mix would be superior to the other if the other was not adjusted. It was necessary to decrease the fines and increase the coarse by approximately 100 pounds per cubic yard to make the mixes equal. Observers thought the change was not necessary but, when the two mixes were tested on the project, there was no difference in placeability or strength.

### U.S. Air Force Research

The Air Force was experiencing premature deterioration of airfield pavements (runways, taxiways, and aprons) as early as one year after completion of the construction. This was a major problem because pieces of the small concrete could be sucked into the aircraft engines. This was especially a problem with the fighter aircraft. The particles are known as foreign object debris or "FOD."

The Civil Engineering Research Foundation (CERF) was asked to intervene and develop a solution. A committee was organized with this author as a member. After discussing many possibilities, the committee agreed that many problems were caused by small spalls at construction and saw joints. Segregated mortar rose to the top of the slab and developed a point of weakness on the surface. ACC suggested that the Air Force track results of poorly performing airfield pavements. Following a six-month investigation, the investigators found performance was dependent upon aggregate grading. After discussing options, combined aggregate grading and the Coarseness Factor Chart were evaluated on new projects to determine their usefulness. The program was tested during construction of a new runway at Ft Rucker, AL. A mixture design was developed and the effects of changes in gradation and their affect on performance were plotted on the CFC. Figure 8 is a compilation of the results provided by James LaFrenz in a private communication.

The mix design point "D" uses less fine aggregate than is generally used because the coarse aggregate was rounded gravel. The batches that were used successfully are identified with an "X". Mix "R" fell in the Trend Bar and was reported to be rocky. Mix "O" was over-sanded and the "S" mixes showed a tendency to spall and slump at the edges. The USAF has used this concept since this work. The latest project to use the method was at an air base in Afghanistan.

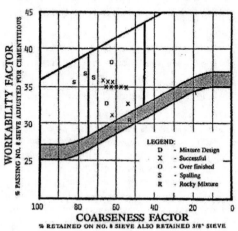

**Figure 8. Compilation of Data from Ft Rucker**

## Proven Results

Figure 9 is the compilation of five slip-formed paving mixes – one a military airfield - that have performed well during construction and during their service life. This information is based upon private communications with equipment suppliers, contractors, and personal experience. It should be noted that a single specification using combined aggregate grading and these graphics can be used across the country and other parts of the world. Emphasis on combined aggregate grading makes it possible to select those locally available materials and use them to meet the needs of a portland cement concrete pavement.

Iowa and Michigan DOTs report concrete strength has been increased with less cement and, at the same time permeability has been reduced to extend durability. The Port Authority of New York and New Jersey applied the technology for construction of an early runway at Newark Airport. A special adaptation is used today by that Authority.

The Federal Aviation Administration is reviewing the final draft of the new FAA P-501 specification for airfields. The technology described here is a mandatory part of the specification.

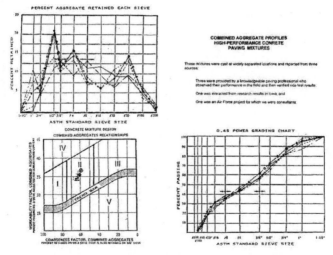

**Figure 9. Paving Mixes with 1-1/2 – Inch Aggregate**

## Conclusions

More is currently being – and will be in the future – learned about optimizing concrete paving mixture technology. In November 2007 there will be an "International Conference on Optimizing Paving Concrete Mixtures and Accelerated Construction and Rehabilitation" where this subject will be discussed. The announcement points to construction of concrete pavements that will perform 40 or more years. Currently, only a 20-year pre maintenance life is expected. And we see process automation as an important part of future concrete mix process control.

A major problem today is that few engineers understand the technology of concrete mixtures. The American Society of Civil Engineers is addressing the problems caused by 4-year engineering education. Engineers can be in the forefront of change for a better, longer lasting highway system. We must answer the age-old question, "Why can't we build them like they use to?" We can build them better than they use to!

## References

ASTM C33, *Standard Specifications and Tests for Concrete and Concrete Aggregates*, American Society for Testing and Materials
*Design and Control of Concrete Mixtures*, Second Edition, Portland Cement Association, January 1927

*Information Exchange Issue No. 58*, Texas Highway Department, 1938.

Powers, T.C., *Properties of Fresh Concrete*, John Wiley & Sons, New York, 1968.

Shilstone, J.M., *Saudi Air National Guard Final Report*, Dallas, 1975.

Shilstone, J.M. *Concrete Mixture Optimization*, CONCRETE INTERNATIONAL, American Concrete Institute, June 1990.

Shilstone, J.M. Sr. and Shilstone, J.M. Jr., *Performance-Based Mixtures and Specifications for Today,* CONCRETE INTERNATIONAL, American Concrete Institute, February 2000.

Weymouth, C. A. G., *Effect of Particle Interference in Mortars and Concrete,* ROCK PRODUCTS, February 1933.

# Subject Index

Page number refers to the first page of paper

# Author Index

Page number refers to the first page of paper